LEAN:

2 BOOKS IN 1.
Agile Project Management + Scrum.
A QuickStart Bundle to Lean

By:

Alex Moore

Agile Project Management:

An Easy Step by Step Handbook to Learn Agile Project Management and Make Innovative Products

By:

Alex Moore

Table of Contents

Chapter One

Introduction

Project management is the application of information, skills, tools and techniques to project activities to satisfy the project necessities. in spite of industry, project management has been established to be an important part of a company's potency and its ultimate success. In fact, the organizations using tried project management practices waste twenty-eight less cash and implement projects that are 2.5 times a lot more flourishing. Project management professionals conclude that the definition of a successful project is one that is not solely completed on time and within budget, however one that conjointly delivers expected advantages.

For many years, there has been a traditional method of project management based on the idea that the customer can know and define his or her requirements fully up front. While this is occasionally the case, very often customers either don't know precisely what they want or, more often, "will know it when they see it." Agile project management is a repetitive and progressive approach to delivering necessities throughout the project life cycle. At the core, agile projects ought to exhibit central values and behaviours of trust, flexibility, authorization and collaboration. Agile project management focuses on continuous improvement, scope flexibility, team input, and delivering essential quality merchandise. Agile project management focuses on continuous improvement, scope flexibility, team input, and delivering essential quality merchandise. Agile project management approaches embody scrum as a framework, extreme programming for building in quality upfront, and lean thinking to eliminate waste.

Why does one want agile in project management? Agile could be a philosophy that concentrates on authorized individuals and their interactions and early and constant delivery valuable into an enterprise. Agile has enduring attractiveness and 'proved' itself in software system development. However, though the arguments are compelling, proof that it's additionally beneficial than other approaches remain for the most part anecdotal.

Firstly, a short summary of what agile project management is and the way it differs from a lot of ancient project management approaches.

There are many methodologies which will be accustomed manage an agile project; two of the simplest famed being scrum and Lean. an agile project's shaping characteristic is that it produces and delivers work in short bursts (or sprints) of anything up to a few weeks. These are repeated to refine the operating deliverable till it meets the client's needs.

Where traditional project management can establish an in depth set up and detailed needs at the beginning then decide to follow the plan, agile starts work with a rough idea of what's needed and by delivering something in an exceedingly short amount of time, clarifies the wants as the project progresses. These frequent unvaried processes are a core characteristic of an agile project and, as a result of this way of operating, cooperative relationships are established between stakeholders and also the team members delivering the work. Scope must be variable where no detailed needs exist initially, however agile still has processes to make sure that, at every stage, the work to be done is outlined and in-line with client wants. The role of project manager tends to be quite totally different on agile projects (and is usually referred to as the scrum Master or Project Facilitator); it's the team member who deals with issues and handles interruptions to permit the other team members to focus on producing the work.

So agile projects need documentation, reviews and processes even as traditional projects do to satisfy needs, manage prices and schedules, deliver advantages and avoid scope creep; agile merely doesn't place as much stress on extremely detailed documentation and does not expect to totally perceive the wants before work can begin. Instead it emphasises the importance of delivering a working product as one thing tangible for the client that can then be refined till it fulfils the client's wants. The key measure of project progress is this series of operating deliverables. there's clearly a risk to starting work on a project before the extent of that job is absolutely known however this risk is quenched by the speedy delivery of a working product, albeit one that's unlikely to be excellent initially.

Chapter Two

History of Agile Project Management

As against the standard methodologies, agile approach has been introduced as an effort to form software engineering versatile and economical. With 94 of the organizations practicing agile in 2015, it's become a typical of project management. The history of agile are often traced back to 1957: at that point Bernie Dimsdale, John von Neumann, Herb Jacobs, and Gerald Weinberg were exploiting progressive development techniques (which are currently called Agile), building software system for IBM and Motorola. Although, not knowing a way to classify the approach they were active, they all realized clearly that it had been completely different from the Waterfall in many ways. However, the modern agile approach was formally introduced in 2001, once a gaggle of seventeen software development professionals met to debate different project management methodologies. Having a transparent vision of the versatile, light-weight and team-oriented software development approach, they mapped it call at the manifesto for Agile software system Development. aimed toward "uncovering better ways in which of developing software", the manifesto clearly specifies the fundamental principles of the new approach:

A Detailed History of Agile:

Here could be a look into how Agile emerged, how it acquired the label Agile, and where it went from there. It's necessary to take a glance at where Agile software system development came from to get an understanding of where things are at nowadays.

Before 2001: A lot of individuals peg the beginning of Agile software system development, and to some extent Agile generally, to a gathering that occurred in 2001 when the term Agile software system development was coined. However, folks started operating in an Agile fashion before that 2001 meeting. beginning in the nineties, there have been numerous practitioners, either folks operating within organizations developing software products or consultants serving to organizations build software who thought, "You know what? The method we've been building software package simply isn't working for us. We've need to come back up with one thing completely different." These software developers started mixing old and new concepts, and once they found a combination that worked, they created a technique for his or her team to assist them bear in mind the combination of ideas that worked in an exceedingly given scenario. These methodologies emphasized close collaboration between the development team and business stakeholders; frequent delivery of business worth, tight, self-organizing teams; and sensible ways to craft, confirm, and deliver code. The people that created those methodologies patterned that others is also fascinated by obtaining a number of identical advantages they were experiencing, in order that they created frameworks to unfold the ideas to alternative groups in other organizations and contexts. this is often where frameworks like scrum, Extreme Programming, Feature-Driven Development, and Dynamic Systems Development method, among others, began to appear. The unfold of the ideas at this point was very organic, and every one of those completely different approaches began to grow in a very grassroots manner. folks borrowed the initial frameworks and tweaked them with completely different practices so as to create them applicable for his or her own contexts.

2001: There wasn't the same approach of describing these alternative ways to develop software package till a bunch of 17 individuals thought, "We're all doing these completely different approaches to developing software. we tend to need to get along and see wherever there are commonalities in what we're brooding about." The result was a gathering at a ski resort in Snowbird, Utah in 2001. When they got along, they did some sport and conjointly discussed where their approaches to software development had commonalities and variations. There were loads of things that they didn't agree upon, however there have been a number of things that they were able to agree upon, which ended up turning into the manifesto for Agile software Development. the 2 main things the Agile manifesto did was to supply a collection important statements that form the base for Agile software development and to coin the term Agile software development itself. In the months later on, the authors enlarged on the ideas of the Agile manifesto with the 12 Principles Behind the Agile manifesto. Some of the authors, together with Martin Fowler, Dave Thomas, Jim Highsmith, and Bob Martin, wrote up their recollections of writing the Agile manifesto. 16 of the 17 authors met at Agile2011 and shared their recollections of the event and their views on the state of Agile up to that purpose.

Post 2001: After the authors returned from Snowbird, Ward Cunningham posted the Agile manifesto, and later the 12 Principles. Agile Alliance was formally formed in late 2001 as an area for people that are developing software and helping others develop software explore and share ideas and experiences. Teams and organizations began to adopt Agile, led primarily by folks doing the development work in the groups. Gradually, managers of these groups conjointly started introducing Agile approaches in their organizations. As Agile became additional wide well-known, an ecosystem shaped that included the people that were doing Agile software package development and also the folks and organizations who helped them through consulting, training, frameworks, and tools. As the scheme began to grow and Agile ideas began to unfold, some adopters lost sight of the values and principles espoused within the manifesto and corresponding principles. rather than following an "agile" mindset, they instead began insisting that certain practices be done specifically in an exceedingly certain approach.

Organizations that focus entirely on the practices and also the rituals expertise difficulties operating in an Agile fashion. Organizations that are serious regarding living up to the Agile values and principles tend to comprehend the advantages they wanted and notice that operating in an Agile fashion isn't any longer one thing that's new and completely different. Instead, it merely becomes the approach they approach work. Agile Alliance continues to minister resources to assist you adopt Agile practices and improve your ability to develop computer code with lightsomeness. The Agile Alliance web site provides access to those resources together with videos and displays from our conferences, expertise reports, an Agile wordbook, a directory of area people teams and several other resources.

Benefits and Rules of Agile Project Management

Agile may well be a project delivery 'placebo'; operating as a result of those involved need it to. Agile empowers people; builds responsibility, encourages diversity of ideas, permits the early release of advantages, and promotes continuous improvement. It permits choices to be tested and rejected early with feedback loops providing advantages that don't seem to be as evident in waterfall.

In addition, it helps deliver modification once requirements are unsure, helps build client and user engagement by focuses on what's most useful, modifications are progressive improvements which might facilitate support cultural change. Agile will facilitate with deciding as feedback loops help economize, re-invest and realise quick wins.

However, Agile focuses on little progressive changes and therefore the challenge is that the larger image will become lost and create uncertainty amongst stakeholders. Building agreement takes time and challenges several norms and expectations. Resource value is higher; co-locating groups or invest in infrastructure for them to work together remotely. The onus is appeared to shift from the authorised end-user to the empowered project team with a risk that benefits are lost as a result of the project team is focussed on the incorrect things.

A critical governance decision is to select the appropriate approach as part of the project strategy. Level of certainty versus time to market is the balance that needs to be considered when selecting suitable projects to go agile. Organisations have to be realistic: the objective is not agile but good delivery, and a measured assessment of the preferred approach is essential to achieve that goal. This is defined by the project type, its objectives and its environment.

Agile is not a panacea, many practices its principles without knowing. Projects delivering end-user benefits is an agile principle which should also exist using traditional methodologies. Collaborative working will always: improve benefits; speed up delivery, improve quality, satisfy stakeholders and realise efficiencies.

There are 12 basic principles to successfully following an Agile project management development approach. At a ski resort in Snowbird, Utah, 17 software developers reflected on what defined the core principles of agile development methods. Their goal was to uncover better ways of delivering software and to help others do the same.

During that meeting, the Agile Manifesto was born. Comprised of 12 fundamentals, along with four core values, it provides the foundation of agile software development as we know it today.

• Attain customer satisfaction through continuous delivery of software

• Don't be afraid to make changes

- Deliver working software, with a preference to the shorter timescale
- Developers and management must work together
- Build projects around motivated individuals
- Face-to-face interactions are the most efficient & effective modes of communication
- Working software is the primary measure of progress
- Agile processes promote sustainable development
- Continuous attention to technical excellence and good design enhances agility
- Simplicity is essential
- The best architectures, requirements, and designs emerge from self-organizing teams
- Inspect & Adapt

•

Principle 1: Attain customer satisfaction through continuous delivery of software

Software is not built for the sake of building software. It's built to be put to use by an end user to better perform tasks that were previously out of reach, solve a problem, do their job better or more efficiently, etc. But often, the highest priority of software development is forgotten.

So, how can you better align with this principle?

Shorten the distance between requirements gathering and customer feedback by planning less change at a time. This gives you more opportunity to steer the software in a satisfactory direction for the customer.

Principle 2: Don't be afraid to make changes

You can implement changes now -- you don't need to wait for the next system to be built or a system redesign. Agile processes harness change for the customer's competitive advantage.

Shorten the distance between conceiving and implementing an important change. And even if it's late in the development process, don't be afraid to make a shift.

Principle 3: Deliver working software, with a preference to the shorter timescale

Previous development methods were front-loaded with tons of documentation under the guise of completing 100% of the requirements needed for a particular project. But towards the end of the project, the usual result was just that -- lots of documentation, but nothing to show for it.

Agile project management focuses on shortening the distance between planning and delivery. So, the agile methodology focusses more on creating software rather than just planning for it. This gives you the opportunity to improve the efficiency and effectiveness of the work.

Principle 4: Developers and management must work together

This one is crucial, especially because it doesn't come naturally to most people. Co-location between management and developers is usually the best way to handle this. You can also use communication tools for remote workers. It helps the two sides better understand each other and leads to more productive work.

Principle 5: Build projects around motivated individuals

There should be no micromanaging in agile project management. Teams should be self-directed and self-reliant. Make sure you have the proper team in place that you can trust to complete the project's objectives and provide the support and environment to get the job done.

Principle 6: Face-to-face interactions are the most efficient & effective modes of communication

Put simply, you want to shorten the time between a question and its answer. This is another reason why co-location or remote work during the same hours is key in agile project management. When teams work together under the same (virtual) roof, it's much easier to ask questions, make suggestions, and communicate.

Principle 7: Working software is the primary measure of progress

This is the primary metric an agile development team should be judged by: Is the software working correctly? Because if it's not, it doesn't matter how many words have been typed, bugs have been fixed, hours have been worked, etc. A good team needs to produce quality software -- all other measures are pretty much irrelevant if you can't get it working correctly.

Principle 8: Agile processes promote sustainable development

When working on the same project for a VERY long time, burnout can be a common problem among agile software development teams. To prevent this, work should be done in short productive bursts because excessive overtime cannot continue indefinitely without impacting the quality. Focus on choosing the right pace for the team members. Usually, the best pace is one that allows team members to leave the office tired yet satisfied.

Principle 9: Continuous attention to technical excellence and good design enhances agility

Developers shouldn't wait to clean up redundant or confusing code. Code should get better with each iteration. Along agile methodology, the software development team should use scrum tools and take time to review their solution. Doing this during the project saves you way more time than cleaning up code "later" -- which can also mean never.

Principle 10: Simplicity is essential

Keep things simple and minimize the time between comprehension and completion. Avoid doing things that don't matter -- such as the "busy work" that is so prevalent in corporate culture. Keep track of your team, count the hours worked in a fun way by using project management tools.

Principle 11: The best architectures, requirements, and designs emerge from self-organizing teams

A great agile management team takes its own direction. Members don't need to be told what needs to be done -- they attack problems, clear obstacles, and find solutions. It should be a red flag if the project manager has to micromanage.

Principle 12: Inspect & Adapt

This is a crucial principle in agile project management. At regular intervals, the team should reflect on how to become more effective, tune and adjust its behaviour accordingly. If there is a better way of moving a project forward, the team should implement adjustments.

Chapter Three

Traditional To Agile Project Management

In recent years, Agile methodology has become popular with several software package development groups as a result of the enhanced potency it brings regarding. quite a few corporations were unable to style and make desired products within optimum time and price thanks to their use of traditional project management methodology. By going agile they were able to fully remodel their processes and alter the means groups view project management.

Project Management: Project management is the discipline of initiating, planning, executing, controlling, and closing the work of a team to attain specific goals and meet specific success criteria. Regardless of the scope, any project ought to follow a sequence of actions to be controlled and managed. in keeping with the Project Management Institute, a typical project management method includes the subsequent phases:

1. Initiation;

2. Planning;

3. Execution;

4. Performance/Monitoring;

5. Project close.

Used as a roadmap to accomplish specific tasks, these phases define the project management lifecycle. Yet, this structure is just too general. A project typically incorporates a variety of internal stages within each phase. they can vary greatly relying on the scope of work, the team, the business and therefore the project itself.

In attempts to search out a universal approach to managing any project, humanity has developed a major number of Project management techniques. based on the above-described classic framework, traditional methodologies take a bit-by-bit approach to the project execution.

Thus, the project goes through the initiation, planning, execution, monitoring straight to its closure in consecutive stages. typically referred to as linear, this approach includes a number of internal phases that are successive and executed in a chronological order. Applied most typically to the development or manufacturing industry, where very little or no changes are needed at each stage, traditional project management has found its application within the software engineering still. referred to as a waterfall model, it's been a dominant software development methodology since the early 1970s.

Traditional (or waterfall) project management, majority of the times, follows a set sequence

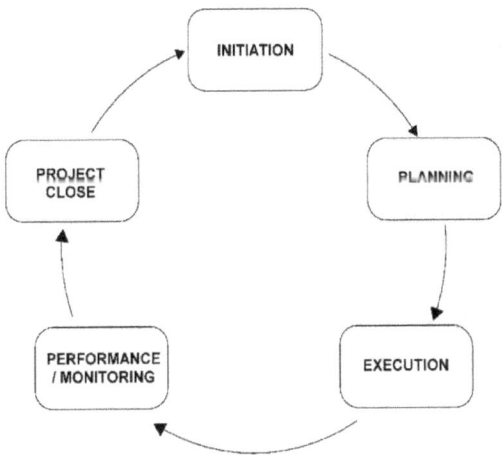

When one process is complete, only then can the next one begins. Typically, this method is more suited for projects that anticipate very few changes from the start to finish. Like manufacturing a billion-dollar Heli carrier. Here requirements are fixed, only cost and time vary. Since there are minimal changes, chances of the budget or time estimation going haywire are comparatively less.

However, not every project can be planned in the same manner. For software development projects, frequent iterations are required, a flexibility which Agile methodology provides. Instead of planning the entire project beforehand, teams focus on quicker iterations and increase efficiency. Although most of the steps involved remain same, it is not necessary that they are carried out in a sequential manner. These steps are broken down into smaller segments known as sprints.

Both Traditional and Agile Project Management have their merits and demerits. It depends on the nature of the product and the circumstances around it which determine the type of method to be used. It is wise to first understand the difference between the two before you consider adopting either one.

Traditional Project Management Model

Traditional project management is a universal practice which includes a set of developed techniques used for planning, estimating, and controlling activities. The aim of those techniques is to reach the desired result on time, within budget, and in accordance with specifications. Traditional project management is mainly used on projects where activities are completed in a sequence and there are rarely any changes.

Traditional Project Management or waterfall model includes a robust stress on planning and specifications development: it's considered to take up to 400th of the project time and budget. Another fundamental principle of this approach is a strict order of the project phases.

Waterfall could be a linear, consecutive design approach where progress flows downwards in one direction—like a waterfall. Originating in the manufacturing and construction industries, its lack of flexibility in design changes within the earlier stages of the development process is because of it turning into exuberantly costlier due to its structured physical environments.

The methodology was first introduced in a piece of writing written in 1970 by Winston W. Royce (although the term 'Waterfall' wasn't used), and emphasizes that you're solely able to move onto the next part of development once the current phase has been completed. The phases are followed in the following order:

1.   Specification requirements

2. Design

3. Implementation

4. Verification

5. Maintenance

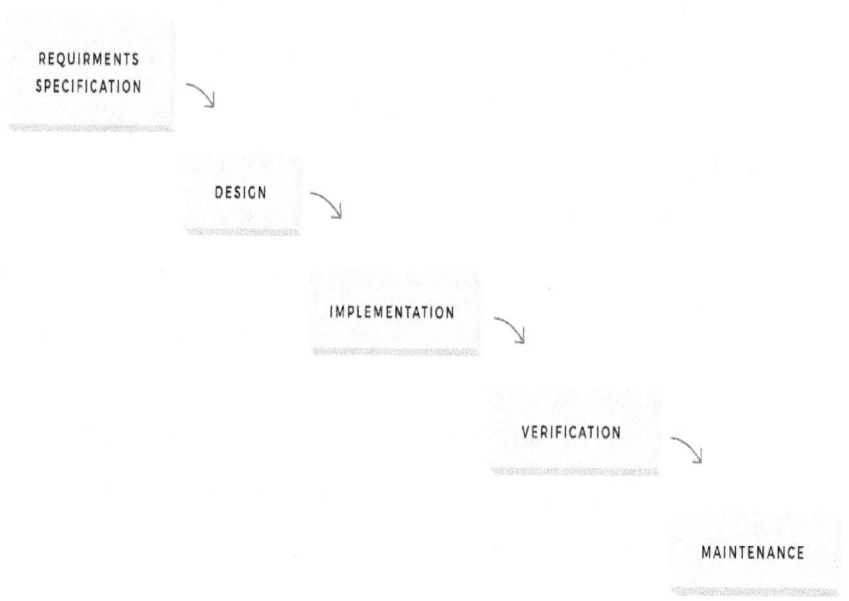

Waterfall may be a project management methodology that stresses the importance of documentation. the thought is that if a employee was to depart during the development process, their replacement will begin wherever they left off by familiarising themselves with the data provided on the documents.

Pre-Agile saw the waterfall methodology being employed for software package development, however there have been several problems because of its non-adaptive design constraints, the dearth of client feedback obtainable throughout the development process, and a delayed testing period. Best fitted to Larger projects that need maintaining rigorous stages and deadlines, or projects that have been done numerous times over where possibilities of surprises throughout the development method are comparatively low.

A new project stage doesn't begin until the previous one is finished. the method works well for clearly defined projects with one deliverable and fixed deadline. waterfall approach needs thorough planning, extensive project documentation and a tight management over the development process.

In theory, this could result in on-time, on-budget delivery, low project risks, and sure final results. However, once applied to the particular software engineering process, waterfall methodology tends to be slow, pricey and inflexible because of the various restrictions. In several cases, its inability to regulate the product to the evolving market necessities usually ends up in a large waste of resources and eventual project failure.

Agile Project Management Model

As opposed to the traditional methodologies, agile approach has been introduced as a trial to make software engineering flexible and efficient. With 94 of the organizations practicing agile in 2015, it's become a typical of project management.

The history of agile can be traced back to 1957: at that time Bernie Dimsdale, John von Neumann, Herb Jacobs, and Gerald Weinberg were using incremental development techniques (which are currently referred to as Agile), building software for IBM and Motorola. Although, not knowing a way to classify the approach they were practicing, all of them realized clearly that it absolutely was totally different from the waterfall in many ways.

However, the contemporary agile approach was formally introduced in 2001, once a bunch of 17 software development professionals met to debate various project management methodologies. Having a transparent vision of the flexible, light-weight and team-oriented software development approach, they mapped it out in the manifesto for Agile software Development. aimed toward "uncovering higher ways of developing software", the manifesto clearly specifies the basic principles of the new approach:

Through this work we have come to value: individuals and interactions over processes and tools operating software over comprehensive documentation client collaboration over contract negotiation Responding to change over following an idea.

Complemented with the Twelve Principles of Agile software, the philosophy has return to be a universal and economical new way to manage projects

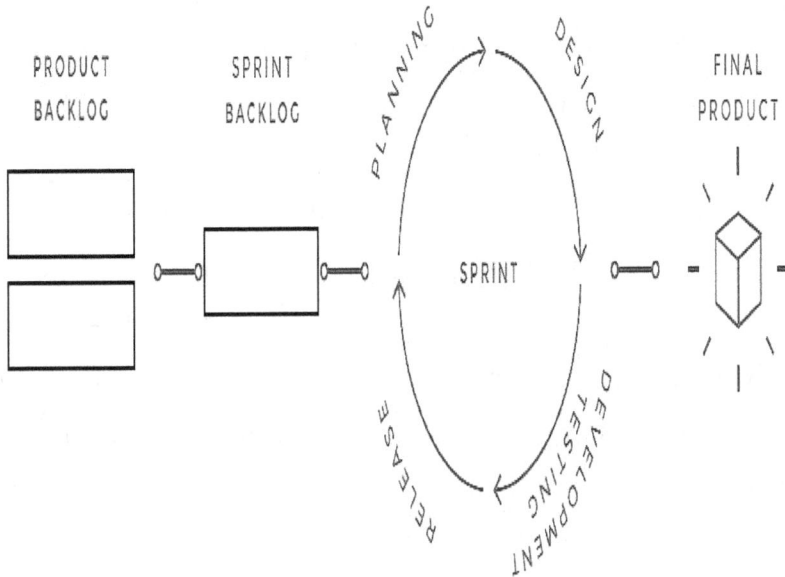

Agile methodologies take a repetitive approach to software development. in contrast to an easy linear waterfall model, agile projects include variety of smaller cycles - sprints. every one of them could be a project in miniature: it's a backlog and consists of design, implementation, testing and deployment stages among the pre-defined scope of work.

Difference between ancient and Agile Project Management

• **Flexibility:**

Traditional project management provides very little to no scope for creating changes to the product. It's a rigid process that solely follows a top-down approach. Once the plan is finalised, managers communicate it to their groups and check that that everybody sticks to that within the absolute best manner. there's plenty of resistance to any modification that's projected because it will disrupt the project schedule.

Agile methodology is a lot of adaptable and offers plenty of flexibility in terms of creating changes to the product. It permits team members to experiment and find out some of the simplest alternatives. they're free to communicate any plan they believe can facilitate to enhance the product additional. Being a feature-based approach, Agile focuses a lot of on obtaining the correct product than follow rigid structures.

• **possession and Transparency:**

In traditional management, possession belongs to the Project Manager. it's the manager's responsibility to arrange and document the whole journey of the product. except managers, solely customers are concerned within the planning stage however once implementation begins their involvement is zero. Since managers hold all the reins of the project, team members typically don't have a say in the results of their efforts or however the project is progressing.

Whereas in Agile methodology, the team members share possession of the project. everybody puts their heads along to come back up with a plan designed to complete the work within the calculable time and price. they're ready to read the progress of the product right from the beginning to its finish. Such transparency plays a very important role in maintaining a productive and extremely engaged work atmosphere.

- **problem solving:**

In case of surprising obstacles, people have to be compelled to escalate the difficulty to their managers. However, approaching your manager each single time isn't a possible possibility. It will cause undue delays and exceed the calculable time limit, except increasing the overall value additionally.

Agile groups have the authority to take choices on their own. they fight to internally solve all problems to avoid dalliance. Being closely concerned in the process, their data helps them to tackle most of the issues that hinder their progress. Unless there's a desire to require extreme choices, team members seldom need to escalate trivial matters to their manager.

- **Checkpoints and monitoring progress:**

Traditional method advocates heavy planning at the analysis and design stage of the project. Their focus is a lot of on streamlining the processes than on the product itself. Once the process is finalised, it's expected that the team can follow it step by step with negligible guidance. Progress is decided when the project is completed. There are not any frequent check-ins unless the manager receives any escalations.

Since there are shorter and faster iterations, Agile methodology encourages team members to have checkpoints at regular intervals. it's straightforward to work out the progress additionally as helps people maintain accountability in their work. In scrum – one in every of the foremost well-liked agile methodologies, groups hold daily stand-ups to catch up on what was the work done yesterday, agenda for the day and if there are any obstacles.

In the dynamic atmosphere that we have a tendency to currently live in, there are only a few things where modification doesn't occur. In such times, sticking to Agile methodology can serve companies far better.

Chapter Four

Why is agile project management necessary?

Agile project management has its disadvantages such as less easy identification of project risks and poor management of resources, and many project teams don't understand how to use agile project management effectively. However, with the fast pace of business change in the 21st century many projects need to be sure they will deliver something that meets client needs at the end of the project and not expend wasted effort refining requirements that will be out of date by the time the end-product is delivered.

Even in business environments that do not change rapidly it can be difficult to fully articulate requirements without seeing a tangible product first so there is still the risk of delivering something that doesn't quite meet the client's needs. That is why agile is becoming increasingly necessary for many different types of projects.

The Multiple and Varied Uses of Agile Project Management

It's no secret that the Agile method is quick turning into the preferred way to manage projects. within the year ahead, Agile is predicted to become even additionally widely used. Organizations, groups and even project management software are more and more responding to a requirement for more accommodative and evolutionary processes. And for good reason. in an exceedingly fast-changing business world that must answer speedy market and technology shifts, Agile delivers. Agile is ideal for any project that needs a series of versions or iterations that require to be reviewed and improved on until the final product is prepared for prime time. for instance, rather than waiting six months for a deliverable that's either blemished or now not meeting the current necessities, Agile permits you to manufacture a primary draft within as very little as two weeks (or less) for immediate feedback—and from here you'll improve upon every version till it's complete.

There are plenty of additional advantages to exploitation Agile. Here are thirteen reasons why groups like yours are exploitation this versatile project management method more and more:

• Agile is evolutionary, giving groups a chance to be told with every new iteration or draft.

• Agile lets groups deliver a prototype and improve upon it with each cycle.

• groups will manage shifting priorities a lot more effectively.

• This quick and versatile method will increase productivity.

• Agile supports regular and collaborative troubleshooting.

• The inherent collaborative nature of Agile improves project visibility.

- Agile helps groups and people effectively prioritize work and options.
- groups will anticipate incoming project changes.
- groups will build quick course corrections based on stakeholder feedback.
- groups can prototype a solution or process for the consecutive version of the project.
- Stakeholders and clients will give feedback because the project evolves—without holding the project up (because the feedback is a component of the process).
- groups get speedy feedback from every version or iteration.
- Empowers project groups to work creatively and effectively.

Agile methodologies address completely customer's wants. throughout the total cycle, user involvement is inspired, providing visibility & transparency, showing the particular progress of projects. As mentioned earlier, Agile method is all about unvarying planning, making it very simple to adapt once some necessities change (if you're employed within the software development trade, i'm positive you recognize what proportion they'll change!). the actual fact that there's continuous planning and feedback through the process means we tend to begin delivering business price from the start of the project. Again, the thought is to deliver business price early within the process, making it easier to lower risks related to development. Let's go through the main advantages of agile project management, point by point.

1. High product quality

In Agile development, testing is integrated during the cycle, which suggests that there are regular checkups to ascertain that the product is functioning during the development. this allows the product owner to create changes if required and therefore the team is aware if there are any problems.

• defining and elaborating requirements just in time so the information of the product features is as relevant as doable.

• Incorporating continuous integration and daily testing into the development process, permitting the development team to deal with problems whereas they're still recent.

• Taking advantage of machine-driven testing tools.

• Conducting sprint retrospectives, permitting the scrum team to incessantly improve processes and work.

• finishing work using the definition of done: developed, tested, integrated, and documented.

• software is developed in progressive, speedy cycles. This ends up in tiny progressive releases with every release building on previous functionality. every release is totally tested to make sure software quality is maintained.

2. Higher client satisfaction

The product owner is often involved, the progress of development has high visibility and flexibility to change is extremely vital. this means engagement and client satisfaction.

• Demonstrating working functionalities to customers in each sprint review.

• Delivering products to market faster and additional typically with each release. The clients get early access to the product during the life cycle.

• Keeping customers involved and engaged throughout projects.

3. increased project control

• **Sprint conferences.**

• **Transparency.**

• **Jira usage (visibility of each step of the project for each parties).**

4. Reduced risks

• **Agile methodologies just about eliminate the probabilities of absolute project failure.**

• **always having a working product, beginning with the very initial sprint, in order that no agile project fails fully.**

• **Developing in sprints, making certain a brief time between initial project investment and either failing quick or knowing that a product or an approach can work.**

• Generating revenue early with self-funding projects, permitting organisations to pay for a project with very little up-front expense.

• Agile provides freedom once new changes have to be compelled to be enforced. they'll be enforced at little or no price due to the frequency of latest increments that are made.

• Adaptation to the client's desires and preferences through the development process. Agile commonly uses user stories with business-focused acceptance criteria to outline product features. By focusing features on the wants of real customers, every feature incrementally delivers worth, not just an IT component. This conjointly provides the chance to beta test software once every iteration, gaining valuable feedback early within the project and providing the power to create changes as needed.

5. Faster ROI

The fact that agile development is unvarying means the options are delivered incrementally, thus advantages are completed early whereas the product is in development process.

• Development starts early.

• A functional 'ready to market' product after few iterations.

• initial Mover Advantage.

- Long delivery cycles are usually a haul for businesses, particularly those in fast-moving markets.

- Agile suggests that quick product releases and skill to measure customer reaction and alter accordingly, keeping you ahead of the competition.

- focusing on Business worth. By permitting the client to work out the priority of features, the team understands what's most significant to the client's business, and may deliver options in the most beneficial order.

Benefits of Agile Methodology

With Agile project management, primary constraints, like time and price, are often unceasingly evaluated. speedy feedback, continuous adaptation and Q&A best practices are designed into teams' schedules, which ensures quality output and a streamlined process. the subsequent represents a number of the other advantages of Agile project management:

- **increased focus on the particular desires of clients**
- **Reduced waste through minimizing resources**
- **increased flexibility enabling groups to simply adapt to change**
- **better management of projects**
- **quicker project turnaround times**
- **quicker detection of product problems or defects**
- **increased frequency of collaboration and feedback**
- **Improved development process**
- **increased success as a result of efforts are more focused**
- **speedy deployment of solutions**

-

Key Aspects of Traditional Project Management Approaches and Agile Development

Traditional Project Management Tools	Agile Project Management Tools
Rigid team members must adapt	Flexible. Can adapt to team members
Abstract	Visual and intuitive
Works best with large teams	Works best with small and medium teams
Reporting is an element of major importance	Constant feedback is important
Works best with teams that share a work space	Can adapt to remote teams
Interest in final result	Interest in each sprint release
Well established quaterly or annual meetings	Daily meetings
Schedule a series of event	Schedule release and product versioning
Works with critical path calculations	Works with burn down charts
Formal management structure	Informal management structure
Dependencies between tasks can be dealt with on the go	Dependencies between tasks must be dealt with as soon as

Quality Within Traditional Quality Management	Agile Development
	possible as they directly affect product releases
Focused on delivering project objectives	Focused on constant improvement of delivered products
Adapts to established requirements	Adapts to customer changing demands
Comprehensive documentations is mandatory	Working software is more important than documentation
Progress is monitored through reports and periodical meetings	Progress is monitored through daily meetings and results
Defines working products criteria	Works with user stories
Reactive response to change	Proactive response to change
Sustainable development	Sustainable development
Complex solutions	Simplicity is essential
Considers that best decisions are made by professionals	Allows teams to self-organize and gives freedom to team members in choosing architecture, requirements and design

Gives power to team leader

Gives power to team members

Chapter Five

Basic quality control

The history of the word quality encapsulates plenty of definitions and aspects as well as case studies. As of lately, quality in all its forms has become a major factor taken into account by all companies and small businesses as well. In this sense, and in order to help companies integrate quality into their everyday business lives, standards, procedures, tools and techniques come to their aid to ensure a standardization and ease of implementation. An interesting aspect emerges however from observing the correlations between the implementation and maintenance of quality throughout a product's lifecycle and the project management process of that product. Another important aspect to factor in is the way that the agile approach and quality merge together to ensure quality products and/or services are delivered.

Quality Management, in a project context, is concerned with having the right processes to ensure both quality product and a quality project. This article describes Traditional Quality Management, Agile versus Traditional Quality Management, Agile Product Quality, Agile Project Quality, Agile Product Quality, Agile Quality Assurance and Control, and Agile Quality Improvement. Quality management is a method for ensuring that all the activities necessary to design, develop and implement a product or service are effective and efficient with respect to the system and its performance. Quality management can be considered to have three main components: quality control, quality assurance and quality improvement. Quality management is focused not only on product quality, but also the means to achieve it.

There are a lot of terms which include the word quality so here is the basic definition of related quality items

Quality policy: An organization's general statement of its beliefs about quality, how quality will come about and its expected result.

Quality management: The application of a quality management system in managing a process to achieve maximum customer satisfaction at the lowest overall cost to the organization while continuing to improve the process.

Quality management system: A formalized system that documents the structure, responsibilities and procedures required to achieve effective quality management.

Quality assurance: All the planned and systematic activities implemented within the quality system that can be demonstrated to provide confidence that a product or service will fulfil requirements for quality.

Quality control: The operational techniques and activities used to fulfil requirements for quality.

Quality plan: A document or set of documents that describe the standards, quality practices, resources and processes pertinent to a specific product, service or project.

Quality audit: A systematic, independent examination and review to determine whether quality activities and related results comply with plans and whether these plans are implemented effectively and are suitable to achieve the objectives.

Project Quality: Notice that Quality Management is concerned with both Product Quality and "the means to achieve it" which Project Perfect: Project Quality Planning calls Project Quality. Project quality is the "things like applying proper project management practices to cost, time, resources, communication etc. It covers managing changes within the project". Interestingly this is pretty close to what is covered by Agile Project Management.

Quality Assurance: Crosby made the analogy that quality assurance is like a person possessing a driver's license. Possessing the driver's licenses provides some confidence that the person is a safe driver

The emphasis of traditional quality assurance is producing a quality plan. A good quality plan, like a driver's license, offers confidence that quality will result. Project Perfect: Project Quality Planning suggests that in a project context the plan should answer the following questions:

1.   What needs to go through a quality check?
2.   What is the most appropriate way to check the quality?
3.   When should it be carried out?
4.   Who should be involved?
5.   What "Quality Materials" should be used?

Other key aspects of quality assurance is producing the quality materials themselves. These include standards, guidelines, checklists, templates, procedures, process, user guides, example documents, and the methodology.

Quality Control: Continuing Crosby's analogy, if quality assurance is like a person possessing a driver's license, then quality control is actually checking that that person is a safe driver. In a project context quality control is about implementing the quality plan, i.e. doing the quality checks described in the plan. Normally just doing the check isn't sufficient and there needs to be proof the quality check took place. In other words the results need to be documented somehow.

Agile Project Quality

Agile is very much concerned about product quality in the sense of "Fitness for use" rather than "conformance to requirements". The first of the Principles Behind the Agile Manifesto is: Our highest priority is to satisfy the customer through early and continuous delivery of valuable software.

It manages to include both satisfied customers and a valuable product, both very quality oriented aspirations. I believe the other element of this principle, continuous delivery, is the key feature of Agile and this principle is very clearly tying continuous delivery to customer satisfaction. Project quality includes "things like applying proper project management practices to cost, time, resources, communication etc. It covers managing changes within the project"

Agile sometimes suffers because people confuse a poor implementation of the method with a limitation of the method itself. A poor implementation will leave out certain key Agile Project Management practices and quality will suffer.

Agile Quality Assurance and Control

Quality Assurance is planning activities to demonstrate quality and Quality Control is implementing those plans. To assure quality a traditional project management project manager would, if they were being thorough, produce a quality plan for the project. Agile project managers don't do this because the Agile process itself provides the quality assurance and control. Agile builds quality in to the product through a combination of practices from Agile Project Monitoring and Control and Agile Project Execution.

Product Owner in the Team: The team is trying to build software that meet's the Product Owner's intent, rather than what they wrote down, so in the Agile world the Product Owner becomes part of the team and guides development. Sometimes the Product Owner cannot devote 100% of their time to the project, but this is a risk that traditional projects also face.

Releasable Software Every Timebox: Each Timebox is meant to result in releasable code. Although meeting the Product Owner's expectations is a priority it is not the only criteria. Releasable software:

1.   Meets the Product Owner's expectations given features asked for and the Timeboxes completed so far

2.   Meets agreed coding standards

3.   Has to best design for the currently implemented features (via refactoring)

4.   Is easily maintainable (via refactoring)

5.   Has been tested to the satisfaction of the team and relevant stakeholders

Unscheduled Product Reviews: Because the Product Owner is part of team they get opportunities to informally review the product. Despite the fact these reviews are informal, teams are encouraged to document the results. Assign a scribe, take notes, type them up, and email them out. I always send the email to the Product Owner and Copy To the Technical Lead and other members of the team working in that area. This is part of Agile Change Management.

Scheduled Product Reviews: The product is formally reviewed in the Timebox Review Meeting at the end of each Timebox. You can schedule more formal reviews during the Timebox, and for longer Timeboxes (say 3 or 4 weeks) this is a good idea. DSDM mandates the results of these reviews are documented. I think this is a good idea but I'm not too tied to review documents. It is recommended to document the reviews in the same as was described in the Unscheduled Product Reviews, i.e. an email to the relevant stakeholders.

Frequent Status Meetings: The team monitors project progress in the Daily Team Meeting. It is also an opportunity to review whether the team is following the agree approach.

Automated Unit Tests: Automated unit tests express the internal behaviours of the software. The total suite of automated unit tests become regression tests and are used to verify that the internals of the software continue to function as designed after subsequent changes. Agile teams aim for 100% pass rate for automated unit tests. More mature teams write the tests before they write the code in Unit Test Driven Development.

Acceptance Tests: In DSDM there are documented quality criteria for all work products, but looser forms of Agile restrict this to the User Stories. User Stories are the high level description of the external behaviours and business rules of your software. Each User Story has at least one acceptance test. The acceptance tests elaborate the brief description provided by the User Story. They define the scope of the story and clarify the Product Owner's intent with concrete examples. This clarifies the Product Owner's intent, points the team in the right direction, and confirm when the intent has been met.

Where possible Acceptance Tests should be automated. As with the automated unit tests, the suite of automated acceptance tests become regression tests, validating that the customer's intent continues to be met by the software after each change to the code. You won't automate all Acceptance Tests – it won't be possible to automate some and others won't be cost-effective to automate. But if you want the test repeated as part of a Regression Test then it is better to automate. If you can write a test so that a person can repeat the steps consistently then you can probably write a automated test and let the computer repeat the steps even more consistently.

More mature teams write the tests before they write the code in Acceptance Test Driven Development.

Test Driven Development: In Test Driven Development the tests become the specification. Because the tests are automated there is no ambiguity, the software either passes the test or it fails.

Regression Testing: A regression test is the repeat of an earlier test. Usually that means Unit Tests and Acceptance tests. Regression tests ensure that changes to the software have not broken good code. My experience is that if regression tests are manual they don't happen. It is with regression testing that the real value of test automation is shown.

Exploratory Testing: Exploratory Testing uses un-scripted tests to quickly identifying new types of problems with the software. Show stoppers (such as system crashes or unhandled errors) are usually fixed immediately. Less serious problems might be deferred. This testing might also reveal new User Stories to be scheduled into later Timeboxes.

Specialist Testing: Extra testing activities like performance testing are scheduled in the same way as User Stories.

Code Review: The Two Pairs of Eyes approach provides a peer review of code to check it follows agreed coding standards, conforms to design guidelines and is easily understood by developers other than the author. This can be either through pair programming or more traditional code reviews/walkthroughs.

Code Metrics: Although not mandated by any of the Agile approaches, some Agile teams, like some traditional teams, collect metrics about the quality of the code. Examples are code-coverage (amount of code covered by unit tests), conformance to maintainability design principles (e.g. Lack of Cohesion of Services, Normalised Distance from the Main Sequence), and language-specific metrics.·

Continuous Integration: Continuous Integration is about maintaining quality all the time, throughout the project. It involves automatically integrating and running a regression test every time somebody does a check in. This is likely to happen several times a day. Running an automated regression test frequently means defects are highlighted soon after they are introduced (i.e. when the build goes Red, i.e. fails). The team's top priority is to get the build Green again.

Informative Workspaces: The Project's Informative Workspace is the primarily place to show data on Project status. Typically it has Burn Down Charts, the Timebox Plan, perhaps the current build status, and anything else that might be a particular concern at the time (e.g. quality metrics such as test coverage). Essentially it is a way to monitor product and project quality on a daily basis.

Scheduled Project Reviews: The Retrospective part of the Timebox Review Meeting is a chance to review the project as whole. Problems are addressed as User Stories in subsequent Timeboxes.

Agile Quality Improvement

One of the Principles Behind the Agile Manifesto is: At regular intervals, the team reflects on how to become more effective, then tunes and adjusts its behavior accordingly.

Retrospectives are the mechanism most Agile teams use for reflection. During a Retrospective the team looks at how well the Timebox went and what they can do different. The high priority changes become User Stories to go into the Release Plan for implementation. Often this is the process to implement new Agile processes.

Chapter six

Tools or Methodology in Agile Project Management

Every agile company has agile project management tools to follow the methodology in a more efficient and strict manner. If you're still doubting which one to choose or if you would like to know which are the leading ones on the market this article is definitely the one for you.

Criteria For Choosing Agile Tools

The best agile tools supply the following most important elements for agile project management. Look at elements outside of their feature set, such as their user interface, their usability (how easy is it to learn?). Also evaluate how much value the tool offers for the price—how its price stacks up against other tools with similar features and functionality.

In terms of features, look for the following in evaluating the best agile tools in this review:

Task management – Kanban or Scrum boards with projects, task lists and everything else that goes with it – from files and discussions to time records and expenses.

Team collaboration – Communicate updates with local and distributed teams, and share task lists, feedback, and assignments

Agile metrics, reporting & analytics – Time tracking and projection, easy-to-understand progress reports for stakeholders, quality assurance, and progress with tools to identify and remedy project obstacles, evaluate performance, and appraise financials

And finally, check for integrations. Ensure that the tool plays well with the right tools. In the case of agile tools, which are often used for developing software, treat integrations with software development and issue management tools with higher priority. However, keep in mind teams in non-development environments won't need this type of integration and would benefit more from integrations with other work apps like Slack, Google Apps, Adobe, etc.

Tools of the trade

The key to success in agile development is to enable flexibility while maintaining organization. The best way to do this is to deploy a set of good tools that help track the project and organize the team's progress. They don't impose strict schedules and roles, but merely make it easier for the developers to self-manage and converge on their goals.

There are dozens of software products designed to help managers set priorities and developers write code that addresses them. Some of these tools are designed to track different forms of development, including projects that are more centrally managed, but they are flexible enough to be used for agile development. Others are built specifically to fit the agile model and nurture as much programmer freedom as possible.

The tools support the project by helping the team identify the requirements and split them into a number of smaller tasks. Then it tracks the programmers as they work collaboratively on the parts. The process is often split up into short cycles that gradually converge on the final result. The cycles alternate between planning sessions and code sprints. Keeping the cycle short and including plenty of developer feedback in the planning lets the team adjust and focus.

A common feature of all these agile tools is a graphic dashboard that reports how the team is progressing and meeting the goals. Some of the more sophisticated tools are integrated with code repositories and continuous integration tools that automatically graph how the new code is evolving. Is the latest code passing tests? Are more features coming online? These questions are all answered on a dashboard that everyone can see. When the team can follow each other's progress visually, they're better able to stay on track.

Another important part of this process is communication. Good agile tools organize the discussion and planning. The developers can focus on each of the features, tasks, or bugs in separate threads. Splitting the discussions up helps the project move forward at the right rate for each section.

Here are the top tools that are forming the foundation teams rely upon to ship code on time or even ahead of schedule.

Source control tools

Git, like some of the other tools here, wasn't built just for agile teams but is still essential. It offers much of the flexibility that teams need to move ahead. The lack of one dominant central repository makes it simpler for different developers to follow different paths and then merge their code later. Git is widely supported, and many teams now use its hosting services to keep their code organized. Many of the other tools in this list take their cues from Git and use the updates to the repository to track and test progress. Other top source control tools include Mercurial, Subversion, and CVS.

Continuous integration tools

Just like Git, continuous integration tools aren't explicitly designed to support agile development, but it would be hard to imagine running a large agile team without their help. The tools automatically add a layer of processing when code is committed, helping to ensure that the team is working smoothly together. The tools have hundreds of plugins for tasks such as creating documentation or compiling statistics. Their most important job is running unit tests that ensure the software is performing correctly after all the new code is added to the stack. Many of the tools in this list also use the results from post-commit testing to determine how quickly the code is meeting goals.

There are a number of good continuous integration tools that play well with agile management systems. Some of the best known tools include Hudson, Jenkins, Travis CI, Strider, and Integrity.

Team management tools

Agile Manager

HP's Agile Manager is built to organize and guide teams from the beginning as they plan and deploy working code through the agile model. At the early stages of the cycle during the release plan, the managers gather the user stories and decide how the teams will attack them. These set the stage for the sprints and deployment.

During each code sprint, the scrum masters and developers record their progress on the user stories and issues. All the progress (or failures) from the build and the unit tests are plotted in charts on a dashboard so the entire team can watch how they're converging on the release.

The tool gathers information directly from major tools such as Jenkins, Git, Bamboo, and Eclipse. To complete the cycle, Agile Manager will push stories and tasks directly to these tools so developers can keep track directly from their favorite IDE.

Active Collab

From juggling tasks to tracking time and generating bills, Active Collab is organized to help software shops deliver code and account for their time. The heart of the system is a list of tasks that can be assigned and tracked from conception to completion. A system-wide calendar helps the team understand and follow everyone's roles. The system checks the amount of time devoted to all the tasks so the team can determine how accurate their estimates are.

The system also supports a collaborative writing tool so everyone can work together on documentation, an essential operation that sets the stage for more agile collaboration later.

The tool can be hosted locally or used through a cloud service.

JIRA Agile

The JIR5A Agile tool adds a layer for agile project management that interacts with the other major tools from Atlassian. The team creates a list of project tasks with a tool called Confluence and then tracks them on an interactive Kanban board that developers can update as they work. The Kanban boards become the center of everyone's focus in planning how to attack the code. JIRA is a tool developed for bug and issue tracking and project management to software and mobile development processes. The JIRA dashboard has many useful functions & features which are able to handle different issues easily. Some of its key features and issues are: issue types, workflows, screens, fields and issue attributes. Some of these features you won't find elsewhere. The dashboard on JIRA can be customized to match your business processes.

The Agile tool is well-integrated with other Atlassian tools. The dashboard updates the moment code is committed to Stash or Bitbucket, Atlassian's Git hosting products. Bamboo (see number three above) offers continuous integration that builds and tests the code before reporting the relative success or failure back to the main JIRA page. Discussions take place through HipChat, which indexes the discussions to the tasks.

Agile Bench

The Agile Bench tool is a hosted platform that emphasizes tracking the work assigned to each individual. The release schedule begins as a backlog of user stories and other enhancements. As they're assigned, the team must gauge both the business impact and the cost of development by assigning an estimate of the complexity of each task in points. The dashboard tracks both of these values so that members can tell who is overloaded and which tasks are the most important. Agile aims to deliver valuable software through close collaboration between all members of the team and stakeholders, welcoming changes to requirements and frequent software releases. This approach means working software can be used much sooner, which reduced confusion around what was actually being delivered. It also emphasises the use of motivated and talented teams of individuals who focus on technical excellence and good design to enhance their agility

The tool is well-integrated with standard Git hosting sites like GitHub or Bitbucket (see number five above), allowing it to make committed code with tasks. If your project needs more, there's also an open API that can integrate the project information with any other system. Popular Alternatives to Agile Bench: Slack, SimplifyEm Property Management, Basecamp, Asana, ClickUp, WorkZone Project Management, Hubstaff, Smartsheet, EclipsePPM, monday.co.

Pivotal Tracker

Pivotal Tracker is a project-planning tool for software development teams. It will help to visualize your projects in the form of stories or virtual cards, break down projects into manageable chunks, have conversations with clients about deliverables and scope. Tracker can divide stories into future iterations, learning from a team natural pace of work. It can accurately predict the estimations and project's completion. Tracker has a transparent team view of priorities to help each member's objectives. Pivotal Tracker encourages a practical agile software development process.

Pivotal Tracker is a straightforward project-planning tool that helps software development teams form realistic expectations about when work might be completed based on the team's ongoing performance. Tracker visualizes your projects in the form of stories (virtual cards) moving through your workflow, encouraging you to break down projects into manageable chunks and have important conversations about deliverables and scope. As your team estimates and prioritizes those stories, Tracker divides them into future iterations, learning from your team's natural pace of work to accurately predict when you will complete future work. Tracker's transparent team view of priorities means that everyone knows what needs to be done, what is being done, and when it will be completed. Tracker's agile philosophy not only helps your team keep pace and plan work, but adjust and change course when the unexpected happens, so your team can deliver earlier and more consistently.

Pivotal Tracker is just one of a constellation of tools from Pivotal Labs created to support agile development. The core of the project is a page that lists the tasks that are often expressed as stories.

Team members can rank the complexity with points, and the tool will track how many tasks are being finished each day. The constellation includes Whiteboard for team-wide discussions, Project Monitor for displaying the status of the build, and Sprout, a configuration tool.

Telerik TeamPulse

Teampulse is an on-premise agile project management software that aims to improve software development processes. In Agile project management, Teampulse enables users to manage requiremnts and bugs. It also hel[s in planning the release of products and in tracking work progress while keeping the project team in cosntant communication and collaboration. It does so through a set of intuitive features that enhances efficiency.

Telerik is known for its numerous frameworks for creating apps for the mobile marketplace. They've bundled much of that experience from creating their own code into TeamPulse, a tool they use to track projects. The main screen displays a page full of tasks that need to be completed and follows the team as it progresses. The menus offer configuration options and a wide variety of reports showing how the project is evolving toward completion. It also works with Telerik's other tools for building and testing code. Telerik allows you to improve our process from requirements gathering through planning, managing and monitoring, resulting in smoother deliveries.

VersionOne

VersionOne is a formidable Agile management solution that is both comprehensive and versatile and developed for teams and projects of various scope and size. It is a compact platform that delivers outstanding performance in terms of managing and tracking multiple teams, tasks, and projects. Simple to use and highly scalable, VersionOne continues to excel in the Agile market, highly used and recommended by businesses of any shape or size.

VersionOne is structured to support other agile software methodologies, including Kanban, Hybrid, Scrum, Lean, SAF, and XP enabling companies to scale agile quicker, simpler, and smarter. Right from planning your portfolio and program level initiatives to tracking and offering value to your customers, the software empowers delivery and reduces time to market. Having an end-to-end visibility of your project's progress and performance equips you with comprehensive insights needed to make data-driven decisions for new plans, changes, and potential issues.

When a large enterprise embraces agile development, they need a tool that's customized to juggle multiple teams working on multiple initiatives because eventually they'll need to work together. VersionOne is designed to organize all the groups involved in development across an enterprise by providing a stable communication platform where everyone can plan the initiatives and create persistent documentation.

The tool embraces Kanban boards for following ideas and stories through the process until they're turned into working code. The system tracks all sprints and organizes the retrospective analysis so the team can start the cycle again.

Additionally, the openAgile API makes it possible to integrate Version One with other packages.

Planbox

Planbox is the pioneering provider of cloud-based AI-Powered Agile Innovation Management solutions – from creative ideas to winning projects. Our mission is to help organizations thrive by transforming the culture of agile work, continuous innovation, and creativity across the entire organization. Our family of products includes Collaborative Innovation Management, Team Decision Making, and Work Management applications. Planbox is designed to provide agile innovation tools for everyone, built for companies and teams of all sizes. Planbox is the comprehensive innovation solution trusted by some of the world's most recognized brands, including Blue Cross, Cargill, Caterpillar, Dow Chemical, Exxon Mobil, Honeywell, John Deere, Novartis, Ontario Power Generation, Sun Life Financial, Whirlpool and Verizon, with millions of internal and external users.

Planbox offers four levels of organizational power to keep multiple teams working together toward a common goal. At the top are initiatives, which are the biggest and broadest abstraction. They contain projects, which are built on items that, in turn, are filled with tasks. As the team finishes the tasks, Planbox tracks the progress on all these levels and produces reports for all stakeholders. One clever feature lets you loop in customers so they can voice their opinion before the code is set in stone. The time tracking feature lets everyone compare the time they spend on an item with the estimate of how long it was thought to take. The tool integrates with Github (see number six above) for code storage, Zendesk for tracking customer satisfaction, UserVoice for bug tracking, and many more.

LeanKit

LeanKit is a cloud-based visual management tool based on a Kanban-style platform that caters to businesses of all sizes across various industry verticals and helps them to implement lean principles, practices and work methodologies across business functions. LeanKit allows organizations to connect project boards at the team and project level and provides users with project visibility. Users can assess project status and manage project dependencies. It also helps users to visualize workflow process and receive real-time updates about project activities and the status of different tasks.

LeanKit also features a reporting and analytics module with metrics such as flow, quality, throughput and lead time. Users can generate custom reports that help them to spot trends and make business decisions. The solution supports integration with various third-party systems such as JIRA, Pivotal Tracker, Salesforce, Zendesk and more.

LeanKit is a visual project delivery tool that enables teams of all types and across all levels of the organization to apply Lean management principles to their work. LeanKit aims to imitate the conference room whiteboards where most projects begin. It lets all team members post virtual notes or cards that represent all the tasks, user stories, or bugs that must be addressed. As the team finishes them, the board updates faster than any whiteboard. The software also allows multiple teams to work together in separate spaces while still coordinating their interactions.

Axosoft

Another widely used Agile Project Management software solution that can be used for bug tracking is Axosoft. Mainly, it is used by software and application developers that are keen of Scrum framework. It has a rich set of tools that every developer needs to ensure that they create and deliver fully functional, bug-free software on schedule. Axosoft's project tool tracks the project in three different ways. The Release Planner offers a tabular view of the different tasks, bugs, and user stories. Developers drag and drop the different entries to assign them and mark them as finished. The burndown charts show graphically how quickly the team is converging on its goal. The projected ship date is displayed prominently to keep everyone on track. The planning is also done Kanban-style using the card view, where each card represents one task.

With Axosoft, developers can create viable plans for development, plot the steps of the process, collaborate effectively and seamlessly, identify issues and resolve them on time prior to delivery. Everything is centralized, ensuring transparency and that everyone is on the same page; supporting the feedback and dialogues option with a customer. Planning with Axosoft should be more effective as the software platform allows to gather all the details and specific information to create the right product backlog. This definitely makes the planning process easy, from creating the steps, scheduling the release, managing the versions and sprints all the way to completion. With the Daily Scrum Mode, project managers, Scrum Masters and other members of the team can see who is assigned to what task and how is one progressing. One useful feature is the customer portal that makes it possible for customers to weigh in on the development process by requesting features, giving feedback on designs, or testing new code.

Agilean

Agilean is an AI and NLP-based SaaS Enterprise workflow automation and management solution that caters specifically to small and medium IT companies. Agilean helps set and automate your Kanban processes under a couple of minutes from the ground up from a selection of over fifty inbuilt templates. Agilean is fully customizable and straightforward for ease of use. Agilean is designed to streamline workflows but also enhance existing assignments based on the specifications of the organization or client. It accelerates and improves user capabilities for project planning, execution, monitoring, control and continuous learning for a variety of software and other industry vertical projects.

Agilean Features include Single dashboard for all projects, visual and clear overview of tasks, drag and drop tasks between columns easily, limit work in progress to be more efficient, horizontal Swimlanes, import Boards, tasks, subtasks, attachments, and comments, automated actions, Gantt Charts, backlog management, create, edit, and move Cards, file attachments, work item breakdown, multi-team work distribution, custom fields, lead and cycle time, work distribution, assigned user, cumulative flow and burn-up, burndown, advanced reporting, work-in-process (WIP) limits, WIP violations and override, map the process, identify impediments, perform the average risk of the project, share response plan and strategy, monitor, control, and close the impediments, schedule meetings and set agendas and simplify meeting minutes, automated follow up on all action points, and real-time collaboration.

With Agilean, the workflow can be constantly improved by removing bottlenecks and time-wasting processes. The dashboard allows a clear view that enables efficiency and a simple way to keep clients happy by responding quickly to their needs and concerns.

It is a SaaS enterprise workflow automation and project management software solution that is basically created to be used by small-medium IT enterprises. The main features of Agilean include project planning, execution, monitor, impediments and response plan, stand up meeting automation, release management, retrospective analysis, and visualized reports.

Wrike

Another great one from the list of the best agile project management tools is Wrike that is one of the best in terms of integrating email with project management, having main features inside. It is built to scale and drive results by giving you the flexibility you might need to manage multiple projects and teams at one place. Along the Agile process, you going to get the accurate, up-to-date information and you can insert & add any important information inside. Planning should be easier, there will be always accurate information and real-time reports and analytics that going to save your time and help in analysing the situation. This is for sure one of the resources that your team might do like to use daily: customization supportive & collaboration tools and many other things that will keep your team focused. The flexibility provided by Wrike enables multifunctional groups to collaborate and get things done effectively from a single location. The service allows you to schedule, prioritize, discuss, and keep track of both work and progress in real time — all with just a few clicks of the mouse.

Wrike has been the project management software of choice for many Fortune 500 companies, such as, Google, Stanford University, Adobe, HTC, and EA Sports to mention a few.

Trello

I guess many of you already know about Trello, one of the most used and well-known project management applications. It has both free and premium accounts that give you a great chance to use most of the common functions. The structure of Trello is based on the Kanban methodology. All the projects are represented by boards, that contain lists. Every list has progressive cards that you make as drag-and-drop. Users that are related to the board can be assigned to said cards. Also, it has many nice, small but not less useful features I would like to indicate: writing comments, inserting attachments, notes, due dates, checklists, coloured labels, integration with other apps, etc. Additionally, Trello is supported by all mobile platforms. What I also like about Trello is that this tool can be used both for work, like we do it in Apiumhub, and personal processes. Trello has a variety of work and personal uses including real estate management, software project management, school bulletin boards, lesson planning, accounting, web design, gaming and law office case management. A rich API as well as email-in capability enables integration with enterprise systems, or with cloud-based integration services like IFTTT and Zapier.

Kanbanize

Kanbanize is a Kanban software for Agile project management that brings full transparency within both individual team workflows as well as across the entire organization. The tool is successfully adopted by a number of industries including Product Development, IT Operations, Marketing and Advertising, Legal and Financial services, etc. Kanbanize is the go-to solution for teams and companies looking to better organize their work, manage multiple projects, track progress and make their work processes more efficient. The software supports highly customizable Kanban boards that allow you to adapt to frequently changing requirements. There, consumers can use timelines to plan their initiatives with greater agility, break them down into manageable tasks, create multiple workflows for cross-functional teams and track overall aging work in progress. The system also supports a powerful analytics module that allows project managers to measure different types of metrics such as lead time, cycle time, team throughput, etc., so they can continuously improve their work processes and make them more predictable.

Backlog

Backlog is an all-in-one online project management tool for task management, version control, and bug tracking. With features like sub tasking, custom issue fields, and Gantt charts, it's easy for teams to define, organize, and track their work. Burndown charts, Git & SVN repositories, and Wikis help developers review, track, release, and document their code. And targeted notifications keep everyone in the loop along the way. Bringing together the organizational benefits of project management with the power and convenience of code management, Backlog enhances team collaboration across organizations large and small.

Assembla

Assembla is a set of tools and services developed to speed up software development and provide support for distributed agile teams. There are 2 platforms offered by Assembla to ensure that teams have the tools and the capabilities to manage, deliver, and maintain not just apps and Agile projects, but websites too. These 2 Assembla products are Assembla Workplaces and Assembla Portfolios. Assembla Workplaces combines various tools and build them around a team list or social activity stream. These include code repositories, management, ticketing and issue management, and collaboration. Assembla Portfolio gives users total control over multiple projects and Team Workspaces. The product comes with a centralized user management feature and reporting plus a branded portal.

Assembla is a combination of cloud-based tasks and code management tools for software developers. The aim of Assembla to move development teams from the typical Scrum agile toward something that is more continuous, distributed, and scalable. Assembla is a provider of Apache Subversion hosting along Git, P4, Dropbox integration, agile task management, team collaboration and project management. With this tool, you can cover all aspects of a project, from ideation to production, as well as to upload large media files, manage code reviews, document your work. Also, if you have any apps you would like to integrate you will easily do it, for example, with Github or Slack.

Asana

Asana is one of the most popular project management software currently available on the market. The robust work management platform serves your teams so they can stay focused on the goals, projects, and daily tasks as you grow your business. To get your works organized, Asana enables you to plan and structure work in a way that's best for you. It handily lets you set priorities and deadlines, share details and assign tasks—all in one place. To stay on track, it allows you to follow projects and tasks through every stage. You know where work stands and can keep everyone aligned on goals.

To help you meet deadlines, the platform lets you create visual project plans to see how every step maps out over time. With it it's much easier to pinpoint risks and eliminate roadblocks. Even when plans change.

Asana is the ultimate task management tool. It allows teams to share, plan, organize, and track the progress of the tasks that each member is working on. It is simple, easy in usage and is free for up to 30 users in a team. As all the previous agile project management software platforms with the main objective in allowing us to manage projects and tasks. What is noticeable is that you don't need to have even an email to use Asana. Each team can create its workplace that will contain projects and project's tasks; each task can have notes, comments, attachments and tags.

This tool can be used as for the small processes and for the giant ones without any limits in the industries or departments.

Binfire

Binfire is an online project management and collaboration tool for decentralized and large teams operating in multiple locations. It helps virtual teams to plan, monitor, and coordinate several projects simultaneously, using a common workspace. Binfire offers all features needed by teams in a single location, so that all files and tasks related to the project can be accessed easily through this one application. Thus, Binfire creates a virtual office space that improves collaboration and communication in the team.

In addition, Binfire is a well-integrated and moderately priced platform, with enterprise pricing adjusted to the needs of different businesses and industries. A 30-days free trial is also available for interested companies to explore the features, and decide which plan works the best for them. Nevertheless, prospective users can count on the company's experienced team to guide them through the process and help them choose, as they can be contacted via phone, email, or 24/7 live support directly on their website. Binfire supports all major project management methodologies including Agile, Waterfall and Hybrid Project Management. It provides real-time collaboration with such features like an interactive whiteboard, message board, burndown charts, project folders, collaborative PDF mark-up, real-time notifications, status updates and much more. In the task management, you can find issue management, bug tracking and document collaboration sections.

Drag

Drag transforms your Gmail into organized Task Lists. It's a free Chrome extension that turns your inbox into a manageable workspace (just like Trello, but for Gmail). Now, many well-known companies like Uber, Airbnb, Netflix, Spotify and others use it to improve the efficiency of email management. On the 18th June they are launching Drag Team! Now, not only individuals can organize their inbox in Trello-style boards, but also teams can collaborate on emails, right from inside Gmail. This means that teams working together in a project or managing accounts such as sales or support don't need to manage multiple external apps to get things done.

Proggio

Proggio is a project management software that puts a premium on collaboration and teamwork. It is built to handle all sorts of projects, both long-term (highly integrated projects) and short-term (continuous stream of deliveries), simple and complicated. With Proggio, project teams have a clear view of their tasks, schedules, and priorities. Managers can easily plan their steps, monitor their progress, and effectively manage budget, resources, and manpower. The project dashboard can be customized to provide users with all the information they need at a single glance.

Users who are on the go can easily see what's going on, what needs to be done, and make critical decisions as Proggio lets them view their project status, tasks, and more from a mobile device. Proggio brings clarity and innovation to agile project management. Proggio is based on a holistic approach to project management, placing people in the center, not tasks. It drives success through creating a shared purpose, building momentum and staying focused. The application introduces several awesome features: amazing project plan visualization, team collaboration in one easy click, patented automatic analysis and process improvements.

nTask

nTask is a cloud-based task management solution that caters to small businesses and individuals. It provides users with tools that enable collaboration with team members, task management, meeting scheduling and more.

With nTask, users can assign tasks, generate progress reports, set recurring tasks, share files, attach files to tasks and generate checklists. Gantt Charts help users monitor project schedules. The solution also enables users to plan and monitor budgets for different projects, allot resources, define risks and issues and monitor team members' time spent on different tasks. From making checklists to managing projects, collaborating with project teams, scheduling meetings, sharing files and more, nTask lets you do everything using just one tool. Spend less time managing tasks and more time doing things with a simple and easy-to-use task board. Worth trying!

OneDesk

OneDesk is designed to improve team collaboration and encourage participation from third parties, and effortlessly connects developers and clients to a productive environment of cooperation for product development. Installed with a rich set of features, OneDesk can do a lot of things to help users come up with better and more customer-centred products and services, from gathering valuable feedbacks and innovative ideas, delivering top class customer support, to monitoring social media for mentions, customer trends, and current buzz, whether positive or negative

With OneDesk you can manage your projects, support your customers, provide services. This tool makes it easy to make successful projects and deliver them on time. It combines Agile and traditional project management: Gantt charts, scheduling, assignments, discussions, notifications on tasks & issues, time-tracking with timesheets & task timers, reporting, exporting, plan releases, roadmaps and many more!

VivifyScrum

VivifyScrum is an Agile project management tool which can be used both by small teams and large organizations. For collaboration, teams can choose a customizable Scrum board which enables working in Sprints, as well as Product Backlog management and other Scrum practices. Teams that prefer Kanban can choose to collaborate on a Kanban board. High-powered item cards allow for a clear description of tasks, sharing all pertinent files and establishing relations between different tasks. In addition to this, VivifyScrum provides features for managing multiple projects across the organization. Thanks to advanced team management functionalities, a company can ensure that the right people work on the right projects. The tool also has an inbuilt time tracker that automatically creates worklogs for team members. Companies can also invoice their clients straight out of VivifyScrum.

VivifyScrum is a cloud-based agile project management solution that features Scrum and Kanban collaboration boards, team management, invoicing, client management and time management. It is suitable for small agile teams and large organizations in multiple industries. Users can create their virtual organizations and add their team members. After creating projects, it is possible to assign various roles to team members and add their engagements on projects. Projects can be linked to the relevant collaboration boards, where team members can track their work. The Scrum board offers product backlog, multiple active sprints, sprint goal, burndown chart and stats per user.

StoriesOnBoard

Stories Onboard is a visual planning tool for Product Owners, agile teams, based on the "user story mapping" method. It's integrated to JIRA, Trello, GitHub, Azure DevOps, Pivotal Tracker etc. More than 1500 companies use this tool worldwide.

StoriesOnBoard is a tool where you can break down your ambitious goals into tangible pieces. Then you can create a roadmap for reaching your goals by identifying the tasks that move toward them the most. By creating a story map, you will be able to see the big picture any time, thus instead of losing in tiny details you can focus on your goals for reaching them in a timely fashion. You can share you story maps with your remote team members so you will be able to work with them online.

Nuvro

Nuvro is a full-featured project management platform that provides the necessary tools to small online businesses at an affordable price range. It empowers owners and managers with a bird's eye view of the projects and tasks their teams are handling with a smart and visual workload calendar. Necessary details such as deadlines and overdue tasks are also visible from this. As such, leaders can act on issues appropriately and with haste.

Moreover, Nuvro promotes smooth collaboration between teammates. It has a file sharing solution with no limits, a team inbox, and a notes module that allows sharing. Because of this, your staff can help each other do better at their assigned responsibilities as well as complete tasks and projects faster. Efficient project execution often relies on many tasks, subtasks and sometimes sub-subtasks. Nuvro was engineered to make it easy to quickly create and assign these tasks to suitable team members. In Nuvro everything's transparent and everyone's accountable for their share of every project. Spend less time trying to bend current software to meet your needs and more time accomplishing your business goals.

Orangescrum

Orangescrum is a task management, collaboration, and project management software combined, providing project managers and teams with an effective platform to help them perform their functions efficiently and improve productivity. The software helps in centralizing all your projects and tasks as well as managing your resources, from people, processes, and technology so that you are able to finish your project in time and without compromising its quality. Orangescrum is also equipped with time tracking capability to help monitor your team members and how much time they spent on their tasks. It also provides you with real-time analytics on all areas of your projects as well as your members, delivering you a clear picture of your projects and enabling you to identify areas that require your concern and attention.

Orange Scrum has a cloud, cloud self-hosted, and open source enterprise editions. Teams can work anywhere and anytime with its Android and iOS mobile apps.

Orangescrum is a simple and effective Project and Task Management software with Open source and Cloud editions. It allows for Agile Scrum Project Management with intuitive Scrum Board, Sprints, Epics, Story points and Velocity Charts. Shape and execute your projects with robust task management (task groups, sprints, tasks & Subtask & Kanban Boards). Interactive Gantt chart allows for real time task progress monitoring and dependency mapping. Orangescrum with its integrated time, resource & invoicing mgmt. prevents the use of multiple tools thereby helping your teams to stay productive and organized. Real time executive reports dashboard makes up for quick informed decisions by your executives. Thus, Orangescrum has it all to put you in control of your business and stay ahead of your competition. It is widely preferred by Small, Medium and Enterprise users. Orangescrum has been ranked among top Project Management Software list for the year 2018 by Accurate Reviews and awarded with Great User Experience and Rising Star for 2017 by Finance online and Compare camp among others.

Zoho Sprints

Zoho Sprints is an online agile project management solution designed to help agile teams plan their project, track their progress, and deliver the appropriate product on time. The simple and clutter-free tool takes care of keeping timesheets, monitoring the task statuses, preparing meetings, and overviewing the analytics. Given that agile teams function with a core value on responding to change and working with a sense of urgency, having a system that augments this type of operation is vital. Zoho Sprints is a dynamic software that is quick and easy to set up, so you can immediately invite your team members, assign them roles, build a backlog, and commence your sprint.

It is part of the revolutionary suite of software solutions by Zoho designed to help businesses in various aspects of their operations. Hence, you can easily integrate Zoho Sprints with Zoho's other applications in marketing, sales, accounting, customer support, and more using a single login and password.

Zoho Sprint is a clutter-free planning and tracking tool for agile teams. It allows you to create user stories, add estimation points, stay on track with personalized scrum boards, and schedule your review and retrospective meetings from one place. Aside from the "To do", "In progress" and "Done" columns, the Zoho Sprint's Scrum Board will allow you to add your own custom columns so they match your team's unique needs, and by using Timesheets you can give your clients a picture of the time involver or estimate your next sprint more properly. Zoho Sprint can be used as an app for both iOS and Android devices, and it allows you to sign up for free.

ProofHub

ProofHub is a cloud – hosted project management solution that helps you to stay on top of deliverables and deadlines. It has scalable features and pay terms that match the requirements of any business size from small start-ups to large enterprise. The Walnut, CA-based company has attracted big ticket clients, such as, TripAdvisor, Harvard University, and Wipro, since the software was launched in 2011. Likewise, ProofHub was among the Cloudswave Awards 2014 top ten project management software for excellent features and performance. Recently, the company introduced ProofHub Bolt apps for Android and iOS users, which made their project management service available even out of the office.

The software helps managers across the critical phases of a project: from planning to organizing, to managing and delivering outcomes on time. It's a central hub for teams, clients, and contractors to share notes, tasks, knowledge, and discussions for a more efficient collaboration and timely response.

A single platform brings together managers and decision makers to discuss plans, create notes and to-do lists, lay down Gantt charts, and calendar milestones and daily tasks. They can drill down the requirements and deadlines to their respective team members by sharing notes, files, schedules, and timesheets. A

Proofing tool also helps relevant parties to discuss, comment, and finalize a document or file in a single window. This tool does away with the cumbersome practice of shuffling email messages and attachments back and forth to finalize a file like logo designs, floor plans, or site map.

Meantime, the Gantt charts, milestone calendar, timesheet, and reports help managers and team leaders track the project and measure on-going performance. Any issues, obstacles, or bottlenecks are easily identified for quick resolutions. ProofHub also features advanced functions, including: Casper mode privacy control; custom domain; SSL encryption, IP restriction, and daily backup; custom roles; and advanced search.

The software integrates well with popular collaboration and productivity tools such as Google Drive and Dropbox and email servers to enhance the overall collaboration experience. Moreover, ProofHub's web browser platform is compatible with any devices and popular operating systems, including iOS, Android, and Windows.

With ProofHub's work management tool you also gain control with Gantt charts to create the project plans that'll ensure timely completion of the projects, assign Project manager to keep a bird's eye view on project progress, create Custom roles to define access levels for the teams and clients, and Project reports for insights on how the projects are progressing, and how teams and individuals are performing.

GanttPRO

GanttPRO is an online project management solution that utilizes a Gantt Chart approach to help users become more efficient and productive in managing their projects, from the conceptualization phase all the way to the realization and delivery. The software is built upon the combined years of experience and top-grade project management expertise that culminate into a powerful platform that integrates all essential project management tools as well as maximum ease-of-use.

GanttPRO is created with the purpose of simplifying workflows, improving collaboration and communication, and delivering projects within accurate estimates. It makes tracking the progress of your projects a breeze and enables you to share your Gantt charts easily with your colleagues and clients during your presentations, reports or business plans.

GanttPRO is one of the most affordable and robust online Gantt chart software in the market with a very intuitive interface. For task and project planning, it offers a visually appealing Gantt chart timeline where all the tasks, dates, and deadlines are clearly seen. This tool is great for team collaboration. You can describe your requirements, add links to useful resources, write comments, and attach files. If any change in a project occurs, GanttPRO notifies team members immediately with the help of push-notifications and emails. For an easy start, GanttPRO offers ready-made templates that will be useful for professionals from different spheres.

Teambook

Teambook is an efficient project management and collaboration tool especially targeted at small and mid-sized businesses, as well as freelancers and service consultants. Its easy-to-use planner with straight-out-of-the-box functions can at once turn your team leaders into effective project schedulers and managers. You can direct company resources for specific tasks using the simple planner. Likewise, you can see who's working on which activities in real-time and track performances and deliverables with the help of analytics. Using tags or filters to refine metrics, search the fitting resource for specific activities.

Teambook also helps you keep tab of the teams with individual resource homepage, email alerts, and iCal link that syncs with popular calendar apps. You client is also updated immediately with a dedicated link to your project milestones. Furthermore, Teambook integrates with popular tools, such as: Harvest, Zapier, Google Calendar, iCal, and Outlook. You can also use its API to integrate your own application

Many organizations don't need a complex and expensive all-in-one project management tool to manage their team. That's where Teambook comes up as a simple and affordable alternative. Teambook is a dedicated capacity planning and resource scheduling online tool that focuses on improving a team's productivity by taking better planning decisions. Its simple visual planner is designed to ensure a smooth planning experience. It gives a fast overview of your team's occupation, adding a booking (unique or recurring) is a matter of seconds and you can edit or move various bookings at once with the bulk editing mode. Plus, Teambook brings powerful filter and custom tag systems to make sure you can easily allocate the right, available resource to the right task. Also, you can identify productivity issues through the simple reporting feature that gives an overview of your resources' availability, productivity and utilization.

Chapter Seven

Problem Solving Techniques

Agile development focuses majorly on adding value to customers. Agile team work on iteration and at the end of every iteration, customer feedback is evaluated which paves the way for further improvement. All these working on iteration and regular interval changes might create host of problems in agile project management. To resolve these problems there are host of problem solving techniques that needs to be taken care of by the project manager and the agile team. Stopping a problem as soon as recovered is an important recipe of success for any project, be it agile or not. While looking for an agile problem, the agile team working on the project needs to look at different levels.

Process level: In this level, the agile team needs to recognize how they are doing with agile adoption.

Quality and performance level: Based on quality and performance level, the project manager needs to understand individual skills and should think of ideas to do better.

The team dynamics dimension: The agile team needs to understand the requirements to make the agile team work better towards achievement of the goal.

There are various Agile Project solving techniques that can be followed by agile team in order to resolve problems in agile and work effectively towards achievement of the agile project. Few of these agile project solving techniques are discussed below.

Establish yourself as a devil's advocate: Make sure that as a project manager or a leader, you will ask a lot of questions. The questions should be asked in order to help the entire team towards some progressive work along with opening up to alternate ideas.

Be kind, rewind: Ask a lot of questions to your agile team and push them towards solving the problems and at the same time motivate them to generate new ideas.

Don't let them waste more time on meta-problems: Relentlessly, focus on the big problem and not on the smaller or Meta problems. It is tempting to stay on Meta problem as they are smaller but not always helpful.

Ask probing questions: Scrutinize the team and understand what they are thinking by asking a lot of probing questions.

Use reflective listening: Reflective listening is a technique in which the listener repeats or paraphrases what he has just heard. The purpose of this technique is to confirm understanding of customer's feedback and let them know that they are heard.

Avoid injecting your own ideas: Even if you have a better idea, always let your team speak first and try to pick the best idea out of them. This will help your team to become more interactive and motivate them to put forwards their ideas.

Lead them to the answer: Always make a habit of allowing your agile team towards the answer. Even though you know the answer, avoid giving it to them directly. Rather ask questions and give hints and lead them towards the answer.

These are some of the best agile problem solving techniques followed by agile practitioners. To know more about agile problem solving techniques and agile project management, you can join Simplilearn's online agile training and improve your PMI-ACP exam preparation towards attaining a PMI-ACP agile certification. You can also attend agile certification classroom training courses for better insights. Know the upcoming dates of PMI-ACP agile certification training for your city.

As organisations have moved away from top-down decision making approaches as the education and skills of their employees have increased, the techniques used for project management have also evolved. Two major changes in organisational approaches to project management are generally grouped under the banners of "agile" and "lean" methods. An agile method relies upon incremental and iterative completion of goals with a self-managing team. It is often presented in opposition to a "waterfall" process that sequentially gathers requirements, completes a design, and then builds a final product.

Hirotaka Takeuchi and Ikujiro Nonaka proposed the core agile concept of iterative, continuous delivery in 1986. They are acknowledged as the inspiration for Scrum, a popular methodology for delivering IT projects today. Co-created by Ken Schwaber, Jeff Sutherland, John Scumniotales and Jeff McKenna, the term "Scrum" is often used interchangeably with "agile". Properly speaking, "Scrum" is a specific methodology whereas "agile" can be any technique that focuses on iterative delivery and empowerment. Agile primarily focuses on efficiently segmenting the business processing cycle of the problem-solving pattern into "chunks" that can be executed in parallel.

A lean method is one that aims to provide "perfect value to the customer through a perfect value creation process that has zero waste". Taiichi Ohno, developer of Kanban at Toyota, came up with the foundations of just-in-time planning and delivery mechanics such as Kanban that are at the core of lean.

In contrast to agile, which splits up work, lean focuses on continuous improvement of cross-functional teams and end-to-end process management. Lean places a far greater emphasis on the knowledge processing cycle of problem solving, continuously seeking new solutions to the status quo.

It is interesting to note the recent emergence of "Lean Six Sigma" advocates; this marks the reclamation of Lean philosophy by top-down management, instead of the more radical Toyota perspective of empowered, trusted, and self-managing teams[6]. But more importantly, following these rules places a relentless focus on discovering new problems, and in continuously building and enhancing the shared context required to effectively solve these problems.

The benefits of using problem solving methods to increase performance and adaptability have been repeatedly demonstrated. In fact, recent research suggests that the process of defining a problem as a series of small projects is often more important than picking a particular execution methodology

The effective problem-solving steps are the following:

•	Identifying the issues: this entails being clear about what the problem is.

•	Have an understanding of everyone's interest: this bring about the unity of purpose of all and sundry in the task of solving the problem. Allow everyone to make his or her contributions in the process of finding a solution to the problem.

•	List the possible solutions: this requires time for brainstorming in order to creatively list the possible solution.

•	Evaluation of the options: here, you will subject the options to a thorough evaluation.

- Selection of useful options.

- Documentation of the agreement: it entails writing down what is agreed upon by the stakeholders.

- Agree on contingencies, monitoring, and evaluation: this is done because there could be a change in condition. You should strive to make contingency agreements about foreseeable future circumstances, how you will monitor compliance and follow-through, and create opportunities for the evaluation of the agreements and their implementation.

Problem-Solving Strategies

A problem-solving strategy is a plan of action used in finding solutions to a problem. In solving a problem, different strategies have different action plans with them. The problem-solving strategies are:

Trial and error: this involves the use of different solutions until you are able to find a solution to the problem.

Algorithm: this is a problem-strategy that offers step-by-step instructions on how the problem can be solved to attain the desired outcome.

Heuristic: it is a general problem-solving framework. It includes the use of the "rule of thumb" which saves you time and energy when making a decision.

How to Solve Problems

In reality, the way you solve your personal or organizational problem will determine your success and happiness. The process of solving a problem includes defining the problem and breaking it into smaller pieces. Also, decide on how to approach the problem.

• You should endeavour to creatively approach the problem by working with other people in order to approach the problem from a different perspective. The following steps help you in solving a problem:

• Define the problem

• Make important decisions first: here, you will recognize the decisions that you need to make and how they will contribute to the task of solving the problem.

• Simplify the problem: this entails breaking the problem into smaller bits.

• Make an outline of what you know and what you don't know

• Be anticipatory about future outcomes: this assists you in coming up with different plans for different problems in the future.

• Allocate resources.

It is crucial for business organizations to have a problem-solving method and model in their workplace because having it enables them to have business problem-solving tools in their arsenal. Problem-solving method and models are used to address the challenges that arise in the workplace. It allows them to solve complex problems confronting their teams with a shared collaborative and systematic approach.

Also, the stakeholders that are involved in running an organization must have creative problem-solving skills at their disposal. This will go a long way in assisting them in confronting their business challenges.

A typical creative problem-solving example was used by XYZ Company, a vegetable oil company. The company was struggling to expand its business operations but the management members were able to creatively solve this problem by selling the shares of the company. With this, the company was able to raise funds required for its expansion.

Problem-solving and decision-making are vital skills for business and life. No doubt, problem-solving involves decision-making and decision-making is crucial for management and leadership. Hence, it is very paramount for business organizations and individuals to have good problem-solving and decision-making skills in order to fulfill their purpose. They also aid in gathering every fact thereby enhancing the understanding of the causes of the problems.

Work problem solving is the method of processing solutions to the work problem of a firm. It involves the collaborative efforts of the workers to solve problems and by doing so; the workers will contribute properly to the productivity of the organization.

A problem-solving flowchart is a diagram that gives a description of how an organization solves its problems. It gives a visualization of the process used in solving organizational problems. It can be used as a guide and reference for the future thereby saving the organizations a lot of time and energy.

Also, a problem-solving flowchart ensures teamwork and collaboration among the workers in confronting challenges they are faced with.

Problem management techniques enable an organization to be responsible for managing the lifecycle of all problems that could arise in an organization. It allows them to prevent problems and resulting incidents from happening, to estimate recurring incidents, and to minimize the impact of incidents that cannot be prevented.

Problem-solving in management is a key feature that any organization must have because it helps in identifying, analysing, and solving the problems that could hinder the growth of the business of the organization.

Also, it is crucial for a business firm to have in place problem-solving leadership that inspires and motivates the workforce to be creative in solving the problems that they will be faced with. Furthermore, problem-solving leadership creates an environment that allows the workers to opine their views on the problems of their organizations.

Agile Team Motivation

If a team want to achieve something they will find a way to make it happen. Conversely if the team don't want to achieve a goal, there is no way they will. Therefore, the single biggest factor which contributes to productivity is the motivation or morale of a team. Treating employees like volunteers means engaging with them and making sure they want to do what they have been asked to do. It would be better yet let them select what task they do Celebrate the success of every launch/release, even if only in a small way, but draw attention to the fact that the team were successful. Allow everyone to feel great about what they have achieved. Praise team members and offer some gold cards like compensatory off.

Traditional thinking is that you need to use a combination of rewards and punishment to motivate people who would otherwise be unmotivated.

Agile Failure Modes

Most of time team is struggling to get enough shared understanding to deliver a working tested increment of product every couple of weeks. With Scrum, we know we need clarity, accountability, and the ability to measure progress on frequent intervals, Once Scrum starts to go mainstream, all people remember the rules but they forgot the meaning behind the rules. Common failure areas are:

Culture doesn't support change - Try to keep cross-organizational uniformity and use PMO as enforcers

Lack of retrospectives Action items generally get lost, hence they should be tracked in system with due date assigned, and should be tracked in reports

Lack of collaboration in planning – Lack of communication and hence lack of collaboration

PMO or scrum master should force to sit in one common room during planning week, where they are forced to collaborate and communicate

Tsunami of technical debt - Stop and clear, one should not process till WIP or technical debt reach to level where it can be controlled

Chapter Eight

Conclusion

Software development methodologies have advanced since business requirements became more demanding. Agile methodologies came into existence after the need for a light way to do software development in order to accommodate changing requirements environment. Agile methodologies rules and practices require communication between the developer and the customer. Under pressure to stick and adhere to the Agile Methodologies principles and best practices, developers must be ready for any change at any time, while also having to maximize Stakeholders Investments. The main aim of agile methodologies is to deliver what is needed when it is needed and nothing more.

Agile Methodologies include a set of software development approaches. They have some variations, but still they share the same basic concepts. The main agile methodologies that are being used today are Extreme Programming (XP), Agile Modeling, and SCRUM. Extreme Programming (XP) is the coding of what the customer specifies, and the testing of that code. Agile Modeling defines a collection of values, principles, and practices which describe how to streamline modeling and documentation efforts. SCRUM is an Agile Methodology framework structured to support complex product development. Scrum consists of Scrum Teams and their associated roles, events, artifacts, and rules. Each component within the framework serves a specific purpose and is essential to Scrum's success and usage. Scrum is a simple low overhead process for managing and tracking software development. It attempts to control this 'chaordic' process using a project management framework that involves requirements gathering, design and programming.

Agile methodologies are not best suited for all projects. When communication between the developer and the customer is difficult, or when the development team does not have experienced developers, Agile Methodologies will not give the best results. These methodologies exhibit optimum results when there is a strong communication between the developer and the customer, and the development team compromises skilled team members. When there is a chance for misunderstanding the exact customer requirements, or when the deadlines and budgets are tight, then Agile methodologies are the optimum approach for a solution.

Agile is a way of thinking about how a software development can be managed. Regardless of the exact frameworks and techniques they use, 98% companies have realized success from Agile projects. Higher speed, flexibility, and productivity achieved through such approaches are the key drivers which motivate more and more organizations to switch to Agile. Software engineering, being an extremely fast-paced industry, calls for flexibility and responsiveness in every aspect of project development. Agile frameworks allow for delivering cutting-edge products and cultivating innovative experiences while keeping the product in sync with the market trends and user requirements.

However, there is always a place for diversity. Depending on your business requirements and goals, you might still benefit from using the Waterfall model or the combination of the two. Specific challenges with using an Agile Method can be offset by adding back some formality. Agile Methods offer software project managers an alternative developement and management methodology that provides good support for projects with ill defined or rapidly changing requiremnts. Even on project that are questionable for the application of the agie method, underlying agile principles may still be effective. Project Managers should consider its usage for such projects assuming that they have a team capable of using it and can implement the required processes.

Scrum

A Step by Step

Pocket Guide to

Make Twice the

Work in Half the

Time with Scrum

By

Alex Moore

contained within this document, including, but not limited to, — errors, omissions, or inaccuracies.

Contents

Introduction

Scrum: the Onetime revolutionary Strategy that Saves Nine

Jakes started an IT firm two and a half years ago, but the experience almost turned sour and the thought regrettable after the managers were unable to understand the tricks of harnessing the human capital and resources available at their disposal. In a bid to reinvent the company and make the initial vision that drove the project work, the naive Board fired and hired new brains and invested fresh funds.

However, despite the novel ideas and resources brought on, things got worse. Then Jakes wanted to quit but his ambitious mind kept spurring him to reengage the process and bring strategies that might change the dwindling fortune of things.

"What is going on," furious Jakes demanded in one of the meetings with the other members of the board.

"I think we keep hiring professional and experts who work at crossroads in terms of vision and purpose," Stems responded in a rather bold-faced manner.

"Mr Stems, what do you insinuate," Jakes queried.

"Or perhaps he's suggesting there is lack of team work and that employees despite all their expertise, tested experience and in-depth skills aren't working with unity of purpose, Nathan cut in.

"Exactly my point, engineer Nathan, Stems quickly responded. "Clearly, there is lack of team work."

Surprisingly, few months down the line, Dan, an outsource expert in company analysis and project management, sought to help Jakes and his team solve the puzzle. Dan had engaged a new invention called 'Scrum' in his previous projects for many of his numerous clients.

So, he introduced the book "Scrum: The Art of Doing Twice the Work in Half the Time" to Jakes' company and by dint of some miracle, the entire IT firm recorded a huge turnaround. And now, it is competing at the top level with the best 20 IT firms in the region.

CHAPTER 1

AGILE PROJECT MANAGEMENT

Perhaps, the best way to start is to say that you've got to stay agile. Yes, you have to. To be agile means to be active, vibrant, and productive. So, when you first heard the word 'Agile' during your first session in business management and software development conference, what was your first reaction like? Confusion? Bewilderment? Something related to your old school days psychology? I guess, you're not alone in the plane of confusion and guesses.

Guess what? Agile is actually a project management software that seeks to help companies achieve more target within a pretty short time. You must be familiar with the biblical Israelites who wasted so much precious and invaluable number of years traversing from Egypt to the Canaan Land. Historians told us the wandering Israelites spent approximately 40 years for a journey that normally should have taken them 40 days. That's crazily unnecessary? Maybe, that for you is a story that best passes for a myth.

In our ever changing world influenced by the impact of technology, there are quite a number of issues that the corporate environment faces which make it increasingly difficult to vouch for customers' loyalty, address support issues, meet clients' needs or achieve project goals and requirements within timeframe.

Historically, companies have been dynamic in their approach, the drive to keep traditional business processes has been challenged with the evanescent nature of the technology-driven business ecosystem.

Not only that the traditional processes are not fast enough, their complexity and sophistication to deal with emerging issues is quite insignificant. Yet, companies have to respond adequately to the needs of the clients.

Hence, these challenges and the necessity of meeting these needs have given rise to technology experts, focused on project management, product management and software development, to build Agile, a new software development process and models of doing things. Differently.

Responsiveness needs have forced from obscurity and nonexistence the evolution of agile and more efficient tools that would not only improve on traditional Waterfall methodologies but also, and more importantly upstage them in the final analysis.

What's need for Agile?

Many project management and software development enthusiasts have asked the usual question any local would ask when a new sheriff enters the town. The question that has surfaced is: why the need for Agile, after all the beautiful functionality and features our traditional Waterfall offers? If the process I'm currently adopting is working perfectly for me, why transition to new model?

But change is the only thing that is constant and you can be sure that every upgrade on an existing model especially in the IT world is always a disruptive one. Imagine the period Windows XP, Linux and Windows 7 were in vogue and compare the functions and features of Windows 8 and 10 to them. The two cycles are world apart, no doubt.

So, the area of product development and management is no different. The product development that was acceptable some ten years ago could no longer be used for the dynamism that changes have brought.

No doubt, what you consider as first rate yesterday can no longer pass that test today. Ditto, what's today's "fast enough" would not be fast enough for tomorrow's needs and changing requirements. The greatest benefit of the Agile technology is to put companies at that competitive edge where they can deploy the Agile processes to help the keep up with the accelerating rate of change. With Agile process, software companies can develop software quicker and at lower costs relative to what is obtainable using the Waterfall. Hence, Agile gives software developers a competitive advantage in a fast-paced market. That sounds theoretical and abstract. Ok, let's get to the benefits of Agile in concrete terms.

As an upgrade over Waterfall, Agile does not require much or in-depth planning at the initial stage of a project. At every stage, Agile practices and methodologies are open to changes based on feedback from end users.

Agile is to be thought of in terms of a familiar process that is, a series of small waterfalls that have very quick iterations and give fast responsiveness ratio. Agile teams are multi and cross-functional and can work on iterations of a product for a time.

Using Agile, you are able to organize all delivered tasks into a backlog. Work iterations are set in relation to the work value relative to the needs and aspirations of the customer and business. At the end each iteration, the product is always work in progress.

However, prioritization of needs and projects as well as alignment of needs of customers and business is an important job that business stakeholders and developers should sit down and work out.

Getting familiar with Agile Practice

So, what is Agile? By simple definition Agile is a software development process. Agile is an incremental, iterative approach to software development.

As a process, Agile is used to describe an approach of managing projects and general attitude to software development. Interestingly, Agile development has its origins and usage in technology rather than in science. In other words, the method is rooted in practice than academia.

Agile practices and processes are best used both in software development and also in all functional areas of the organizational life. This is why leadership in Agile methodologies encourages teamwork, close customer collaboration, frequent and prompt deliveries of working tools, direct communication, accountability, and ability to respond to change.

Agile was originally develop to give a more change responsive process leaders and software developers. Hence, with Agile, industry managers and IT experts can respond to changes and changing requirements with the process rolling out needed functionality.

Besides, in mind for developing Agile framework and making it an advanced model over Waterfall is the need to respond to a number of questions which businesses and software developers seek to find answer to. Some of those questions include:

- What should good Agile practitioners and teams know and do?
- How do we train developers and teams to become good Agile-compliant users?
- How do we transition from plan-driven Waterfall development to practice-oriented Agile development?
- What available tools to practice Agile development?
- How are those tools to be used to give support to Agile practices?
- How can businesses and companies do to manage projects using various Agile practices and frameworks?
- How can companies, organizational processes, and industry develop an efficiency model that would help towards achieving team-targeted rather than individual member-inclined teamwork?

- How do industries define their task job in terms of the client's mission?

- How should software companies not define their jobs as analysts, designers, programmers, testers, or project managers?

Benefits of Agile Processes

Allows for flexible change

While there is focus on value, change becomes investable to processes. Agile processes allow for real change during iteration. Items on Product Backlog can always be reprioritized and refined to meet the need that is urgent. The team has more control in managing and customizing the Product Backlog, review processes and view project progress. The work needs to be shipped early to get expected Return on investment.

Predictable Costs and Schedule

The fixed nature of the Sprint gives the team a sense of predictability in terms of cost, schedule and result of the work done. The team is able to combine estimated costs before each Sprint, giving the client a sense of what it would cost them to have their project done by the team. In this way, there is seamless and more improved decision-making opportunities and prioritizing on the part of both the team and the client.

It facilitates Transparency

The agile approach actively engages the client from the project's start to finish. The processes involve the clients right the blast of whistle at the iteration planning stage through review sessions, up till new feature builds in the software. Clients are able to see and follow up on a project all through stages and not until the final end of the project. That in itself encourages transparency and all-party involvement.

Focuses on Value

Agile provides the platform for the team to focus on value their software brings to business table. They focus on answering questions: What does the client need? How do we help the business to grow? How does our software help deliver features that give the most value to client's business?

User-focused

Agile processes define product features and descriptions as they help business growth and product acceptance using the user's stories. Focusing on the user's needs and expectations helps the team to deliver real value and not building an IT project for consumers.

During each process, Agile processes give opportunity for users and consumers to evaluate and give feedback to the developers and team through testing after each Sprint.

Improves Quality

Agile encourages projects to be broken into handy units in order to ease the burden of heavy task for the team. The focus is to enhance quality and prompt delivery of development. In that way, project testing and team collaboration become seamless. By creating smaller teams that focus on different part of project, the results become more qualitative. N that way, too, conducting tests reviews throughout the iteration, defects and mismatches can easily be spotted and fixed.

It sets purpose for your team

Agile methodologies focus on creating a value-added and shared sense of ownership and objectives for all team members. Put differently, every member of the team is an owner of the project. What this does is that it gives your team a sense of belonging and purpose rather than creating a false sense of urgency. Interestingly, purposeful teams achieve more and are more productive and efficient than a pack of individuals with no synergy.

CHAPTER 2

AGILE MANIFESTO

Every project as it were is driven by certain core values and short, medium and long-term plans. These values consist of the guiding principles that should inform the actions and tasks of creators and innovators of designs and inventions in any field.

Authors of the Agile project also believe that there are certain ethics and principles which should guide the works and operations of software developers. Harnessing these values help software developers build trust with team and clients, and confidence and competence in themselves.

What is Agile Manifesto?

Agile Manifesto therefore refers to a declaration which expresses four core values and 12 principles software developers can use to grow competence and develop quality IT service to clients.

The understanding is that proper management and maintenance of excellent relationship with clients goes beyond possession of skills and expertise. Agile authors also believe that it is not just sufficient to procure standard IT contract negotiations. There are best practices that can effectively help software developers maximize quality service delivery, especially in an environment where there are multiple service providers in the IT industry.

The manifesto contains four key values and 12 principles. Each of the section provides software developers with new insight into decentralizing the heavy processes that hitherto define software development.

The aim of the Manifesto

The aim of the manifesto, among other things, is to overhaul the entire process of software development. For them, the processes of developing software over the years have been encumbered with a lot of bureaucracy. The processes, too, are unresponsive and uni-dimensional in terms of documentation requirements.

The four key values outlined in the Agile Manifesto seek to promote process in software development that focuses on quality through creation of products that meet the needs and expectations of consumers.

In the same vein, the 12 principles are intended to fashion and backing a software development work environment which targets the customer. In that process, the principles will create an environment that relates to business goals and objectives, no less responding and pivoting quickly to changes and feedbacks from the point of view of the end users.

Hence, the manifesto is seeking to maintain a balance and restore credibility to the word of methodology. The implementation of the Agile manifesto, the authors believe, would be not only to plan but also recognize the limit of planning in an ever changing digital ecosystem.

Furthermore, the authors focus on a software development approach that is committed to creating software incrementally.

The approach will in a way provide Agile users with new versions, or releases of software following brief sprints.

Four values of Agile Manifesto

The Agile Manifesto declares four core values of Agile software development. They include:

- Individuals and interactions over processes and tools
- Working software over comprehensive documentation;
- Customer collaboration over contract negotiation; and
- Responding to change over following a plan.

The 12 principles of Agile Manifesto

The 12 principles declared in the Agile Manifesto include:

1. Customer Satisfaction

The Agile methodologies are designed to meet the needs and expectations of customers through early and continuous delivery of valuable work. It is believed that software developers can only earn trust of customers through prompt and resilient prioritization of their needs.

2. Keeping sizeable workload

The Agile methodologies consider breaking down big task into smaller manageable units that can be completed quickly. Simplicity is the watchword for the Agile Manifesto.

Essentially, all Agile methodologies require the art of maximizing the amount of work not done. They ruthlessly focus on cutting down functionality that does not lend value to the chin process.

3. Self-structured team

Agile recognizes that having the best works, architectures, designs, and delivery requires that the team is properly organized. However, a team does not get structured from without; it has to self-motivate and self-organize in order to deliver best solution.

In that process, a self-organized team becomes cross and multi-functional, identifying potential threats and project issues even before they constitute real impediments to the project.

4. Welcoming Change Requirements

Software developers can be more productive and responsive to customers' needs only if they recognize the necessity of welcoming change, even late in developmental stages of the project. Agile processes and methodologies are designed in such a way that they welcome changing requirements.

The change however, should be harnessed towards customer's competitive advantage. The Agile project discourages despair in the face of change, however tough the changing requirements could present themselves to the developer or project manager.

Reacting to change as fast and excellent as possible gives the developer a leverage to get closer to client's needs, and it is a good signpost of progress.

5. Sustainable Effort

Creating processes that promote and support sustainable efforts and collective development is very essential to fashioning a balanced working environment, and Agile processes and methodologies identify and promote that initiative.

Ideally, Agile is designed in such that it encourages that every member in the chain process- sponsors, developers and end users- should be able to work at constant pace and maintain that indefinitely.

The slogan, 'think, work, and balance' quite fits into the proposal of a sustainable effort. Everyone should deeply be involved in getting the project to its final lap. In this way, there would be quality work done in the same degree that the team is also qualitatively impacted.

Agile strives to maintain a consistent level of activity among members in the chain process. The consistency will translate to a better ability to forecast.

6. Measuring progress

What about measuring progress by the amount of completed work? This is a key aspect of the Agile methodologies. Developing software is one of the key factors in measuring work progress.

The objective of the goal must take precedence over religious following of plan. This is so because the more involved you get in following plans, the more distracted you get away from the real goal of the project.

Agile deemphasizes constant documentation update that does not result from measuring progress but strict obedience to plans.

7. Value-added and update-driven solution

As far as Agile is concerned, technical excellence and good design enhance agility. Hence, it is important to pay close attention to this aspect of the project.

The solution derived from a beautiful design is more valuable and meaningful than having an elegant design that does give a result that will stand the test of time.

Agile processes also believes that what is more so is the solution that is capable and open to constant update that will keep it in the loop of currency.

An elegant design, in Agile reckoning must not only be solution provider but also and more importantly delivers a solution that can maintain its value through update and maintenance cycles.

8. Frequent and fast project Delivery

Agile methodologies focus on delivering frequently. The idea is that frequent delivery of working software helps developers to get faster feedback from end uses. The developers will in turn be able to quickly identify what needs to be changed.

The sooner a developer delivers incremental software, the better for the team. Preference to shorter timescale is fundamental to help spot a wrong turn in the process of developing or communicating with the client.

It would be better in Agile mechanism to find out earlier where an errors lies and promptly fix it than having a complete work required for rework.

9. Working through the project

Project managers and software developers must work together on the project throughout the entire process. Also, it is not out of place to have the customer take part in the process of project delivery.

Working through the project consists in understanding that both the customer and the developer are geared towards achieving the same goal.

10. Direct Conversation

Adopting face-to-face mechanism in the process of communication among team members is as crucial as the overall goal of the team. Team performs most efficiently and effectively if they are able of inculcate the method of conveying information face-to-face.

While being in the same location may be ideal and encouraged, having an osmotic sort of communication where co-location is impossible, can also effectively deliver the same result, if handled properly.

Team leaders must keep everyone in the loop of development via direct communication. Using a third party to deliver message may hamper the entire process and so defeat the goal of the team.

It is important therefore to improve the technical aspect of communication techniques among team.

11. Motivating Team Members

Individuals in a team want to build projects. They want to be part of the trusted with projects. Software developers and project managers must motivate individual team members by providing them with excellent environment that supports them.

Agile processes and methodologies emphasize self-organizing teams who impulsively and without compulsion are able to manage both themselves and the work. The need for a micromanagement of projects may no longer be required.

12. Use After Action Review for Effectiveness

Reflection is one key the Agile Manifesto also declares as one principle that delivers incredible outcome and synergizes team. At regular intervals, team members must meet to deliberate on how they can become more efficient.

It is in the process they can inject new ideas, tune and adjust members' behaviours. The use of After-Action Review helps you improve on the next project for the next client. And it is important to review previously finished projects in order to deliver a better one in the future.

Agile projects come with several ceremonies, among which is the Retrospective. At the end of each Sprint or Iteration, Agile encourages that teams should meet in order to catch and improve behaviors before they start a new project.

Not carrying out the review portends a grave danger and detrimental impact on the project. Since agile is based on transparency, technical excellence, respect, trust and commitment, creating high-performing teams helps value individuals and interactions over processes and tools.

CHAPTER 3

AGILE METHODOLOGIES

Agile methodologies refer to the different models and mechanisms in which the Agile practice can be implemented. The type of methodology to be used depends on the project the team wants to work on.

There are several different frameworks used to execute Agile process. While they take their inspiration and mechanism from the same Agile source, each of the methodologies differ slightly in the manner they are implemented. These include:

1. **Extreme programming**

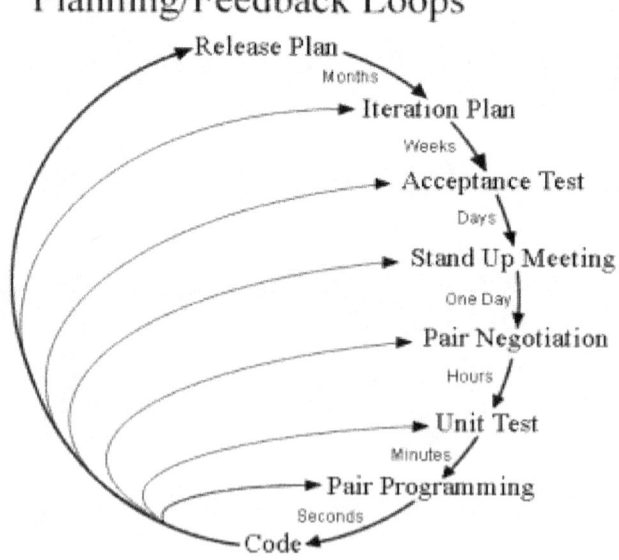

This is one of the methodologies used to implement Agile practice. It is a software development and project management method that intends to enhance quality and responsiveness to change and changing requirements.

It also refers to that agile development methodology that is deployed in software development which allows programmers and software developers to make decisions on the scope of deliveries.

Extreme Programming (XP) remains one of the most popular and yet controversial agile frameworks. For software developers looking to continually and quickly deliver software of high quality, then this highly disciplined methodology is the best o use. One of its features is that Extreme Programming encourages customer participation and each one is involved with the closely knitted team from the planning stage.

The methodology of Extreme Programming operates on four key principles:

1. Simplicity
2. Communication

3. Feedback

4. Courage

In addition to these values, the framework also has 12 backup practices:

1. Plan the game
2. Make releases in small milestones
3. Take customer acceptance tests
4. Make design simple
5. Execute programming in pairs
6. Development should be driven by test
7. Feature again and again
8. Make integration continuous
9. Team ownership of code
10. Set standards for coding
11. Representation
12. Keep workable working speed

2. Kanban

Kanban is framework that operates on the principle of visual implementation of Agile. It is a technique for managing work, with a highlight on just-in-time delivery. Associated with the Kanban framework is the Kanban Board.

The Kanban board refers to a work and workflow visualisation device which provides a summary of the status, progress, and issues related to the work. This model of agile execution encourages mini and unbroken changes to the related system.

Kanban framework operates on the following principles:

- Visual workflow
- Restricting work-in-progress
- Enhancement and management of workflow
- Making explicit policies
- Continuous improvement

Kanban Four Basic Operation Principles

Kanban operates on four basic principles:

1. No setup, no procedure

That sounds raw! That is exactly what it is. Like much of what defines Agile project, Kanban does not have a preset set nor does it prescribe a certain procedure. The framework allows the team to start with it has now; you can always overlay Kanban properties over existing workflow. In this way, you can bring in change.

2. Be ruthless with incremental not sweeping change

Kanban operates with an ideal that suggests accord with incremental, evolutionary change. Since the framework is designed to meet minimal opposition, it encourages continuous incremental and evolutionary changes to your current system. The framework does not encourage sweeping changes because of great resistance capacity.

3. Focus on existing process and role

The Kanban design is one which recognizes that the existing roles, process and responsibilities have values in themselves. Hence,

Kanban doesn't prohibit process; it doesn't prescribe either. Existing process can yield you desired result and generate broader support for your Kanban implementation. But a sudden change can alter setup and progress.

4. Leadership must show at all levels

This principle is not an exclusive operational procedure of the Kanban methodology. Many Agile methodologies including Scrum, RAD also adopt it. It doesn't matter which part of the ladder rung you belong, you should act leadership.

Put simply, you don't have to be a team or executive before you play that leadership role. If you occupy the frontline any team, you are expected to act that quality. Endeavor to show character trait that fosters team spirit and a mindset that encourages continual enhancement to reach team goal.

Contrasting Kanban with Scrum Models

Kanban shares some indistinguishable similarities with the Scrum framework. However, the two have some differences which should not be confused with each other.

	Scrum	Kanban
Cadence	Regular fixed length sprints (that is, 2 weeks)	Continuous flow
Method of Release	At the end of each sprint depending on whether or	Continuous delivery or at the team's discretion

		not the product owner approves it
Roles	Product owner, scrum master, scrum development team	No existing roles. Some teams enlist the help of an agile coach
Key metrics	Velocity	Cycle time
Change process	Teams should strive to not make changes to the sprint forecast during the sprint. Any change can potentially compromise learnings	Change can happen at any time during sprint

3. Rapid Application Development (RAD)

This is one Agile Development methodology which enables software developers and programmers to build solutions with the speed of light by talking directly to end users to meet business requirement.

In simple terms, RAD is less talk, more action because it de-emphasizes strict planning. Although the methodology stresses action over plan, using RAD requires that the developer still takes some steps through the development. The steps in order are figure out requirements, build prototypes, get user feedback, build again, test, and implement.

RAD is cool with your team especially when you need to get some project done quickly. It delivers a working system more quickly than a traditional technique such as Waterfall would do. Also make use of RAD only when you have the budget. This is because it requires you to have a team of highly professional and skilled developers. And you know they would demand some cool cash. Thirdly, you can use Rapid application development when you have an available pool of users and clients who ca reliably test you prototypes.

Why using RAD?

Some of the advantages of using RAD to build your software include that you get a working product more quickly. By that we mean that you can present your work-in-progress in piecemeal allowing your team to put them all together at the end of the whole process.

Also, using allows you to get direct and constant feedback from user. Since you can show your user or client your work in progress, they send to you what you need to remove, add or adjust. You can get their feedback directly and as quick as you want them. That gives you opportunity to improve on your work and implementation becomes easier.

Furthermore, using RAD gives you the latitude to break a large project into smaller units and tasks. There are two things opportunities this offers you. One, when developing a large application, breaking it into smaller unites allows you to form a more specialized team of developers who concentrate on a certain area of the project. The second advantage flows from the first: you can create small wins for your team, allowing you to motivate them. In that way, you have more hands-on deck to help through then entire project.

4. Lean Software Development

Lean software development is one of the Agile iteration processes. It was originally developed by Mary and Tom Poppendieck. Learn is a set of principles that are applicable to software development in order to reduce stress on cost, programming, mismatches and defect rates. The methodology focuses on adding value and giving customer an efficient and value-oriented mechanism t their project.

Seven principles of the Lean Framework

- Waste Removal
- Amplify Learning
- Late Decision Making
- Quick result Delivery
- Empower the team
- Build Integrity
- Integrally envision the whole Application

Benefits of the Lean Framework

1. Lean eliminates roadblocks

In the context of the Learn software development, waste refers to the anything that is capable of reducing the quality of code or hindering time and effort to be spent on coding. Waste could also be defined as any the roadblock to business value deliverables. Example of such bottlenecks include unnecessary code or functionality, delays in programming, blurred requirements, and insufficient testing.

2. Provides customer needs at less cost

The focus of Lean methodology there is to eliminate these hindrances, adopting requisite technologies, and gaining insight into what the customer really needs and expects. One other advantage of the lean framework is that customers can make late but informed decision thus reducing the cost of their investment on the project.

3. Enhances business value

As an iterative development framework, Lean is poised to deliver new applications for enhancements in a quick and smart way. Thus, integrity is built into it to make ensure seamless flow of architecture and system components.

4. Easy to Incorporate

Lean principle also gives organizations the leverage to integrate and achieve continuous improvement as they rapidly introduce and implement changes in their system.

5. Crystal

Crystal Agile framework is one methodology in the development of software that is both lightweight and adaptable. As an approach, Crystal features several agile processes including Clear, Crystal Yellow, Crystal Orange, and other unique methods characteristic of Agile.

There are several factors that drive Crystal processes, which include the size of the team, the criticality of the system, and the priorities of the project the team is undertaking. Crystal method operates with the principle that each project is unique in its own right. Correspondingly, too, the policies and practices that would guide their implementation must also be tailored to their specific requirements and features to be able to meet the need of the customer.

Crystal, like most of the Agile processes, also operates with its unique tenets and principles. In addition to these tenets, Crystal focuses on promoting early and frequent working software releases. Besides, this Agile process also encourages high user engagement, builds adaptability and eliminates distractions and stifling bureaucracy

Crystal principles are:

- Teamwork
- Communication
- Simplicity
- Reflection
- Frequent adjustments
- Improved processes

CHAPTER 4

SPRINT AND SPRINT CYCLE

What is Sprint?

Scrum sprint is a term associated with and used mainly in Agile methodology. It refers to one time-boxed iteration of a continuous cycle in a scrum development. Within a sprint, a team plans, designs and defines the amount of work it wants to complete and made ready for review. In a short common meaning, especially as used in athletics, Sprint could mean a short race at full speed.

Development Teams define a short duration of a sprint, usually between 2 to 4 weeks. Scrum Sprint requires that the team collectively sets the target (technically referred to as Sprint Goal) they want to meet and this is always done in collaboration with the Product Owner. The work plan is itemized in order of priority in the Sprint backlog during Sprint Planning session.

Once the team starts the scrum sprint, team works together to complete planned work effectively and make it ready for review by Product Owner, Scrum Master, stakeholders, customers, end users, among others, by the end of that period.

Before starting a Sprint cycle, a team is expected and mandatorily required to have readied high level User Stories in Product backlog. With the help of Sprint Analytics, Scrum Master and Product Owner can monitor the progress of team work in Sprint in one full glance. The Sprint Analytics helps the Development team, Scrum Master and product Owner to define Sprint Goal and analyze the work done within each Sprint.

The figure below shows an overview of Scrum Flow for one Sprint:

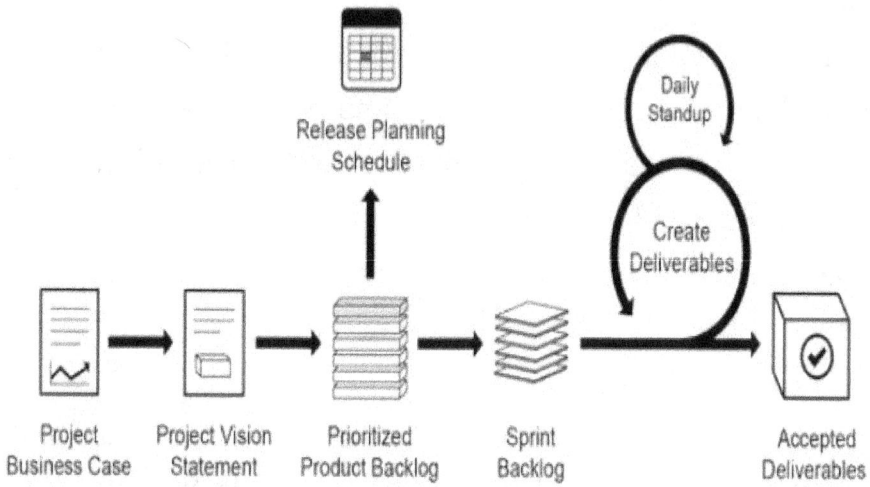

Sprint Cycle

Each Sprint starts with a Sprint Planning Meeting. It is at this coming –together the team considers including high priority User Stories in the Sprint. Generally, a Sprint does not last more than between one and six weeks. During this period, the Scrum Team works on generating potentially shippable Deliverables or product increments.

While the Sprint is underway, the team (which include the Product Owner, Scrum Master, Development Team and stakeholders) also conducts short, highly focused Daily Standup Meetings. This is where team members deliberate on daily progress.

Again, towards the end of the Sprint, the team conducts a Sprint Review as a way of getting stakeholders involved. At this Review, the Scrum Development Team provides demonstrable deliverables of the project they have worked on to the Product Owner and relevant stakeholders.

The purpose of the Review is to have their feedback and remarks on the projects. The Product Owner accepts the Deliverables only if they meet the predefined Acceptance Criteria and Project Requirements. However, the project is rejected and the team is asked to go and rework it to meet the expectations.

After the Review, the Development Team conducts what is called the Sprint Retrospect Meeting where members discuss ways to improve processes and performance in order to deliver better result that they would demonstrate to the rest of the team in the next Sprint Review.

Is there a difference between Sprint and Iteration?

This is one question that a lot of new initiates into the Agile process often like to ask. Especially with the changing trends and advancing technologies, mobile and web applications developers are finding it hard and sometimes difficult to adopt to this changes ad advancements. Other similar include is Sprint just another term for Iteration? Can we have Sprints within Iterations or Iterations within Sprints? What is the interim release to a client before the planned Sprint release data called?

An iteration is the umbrella agile term used for a single development cycle. It is often a commonly used in the IID, that is, Iterative and Incremental Development. Scrum, on the other hand, is a specialized agile framework or a specialized Incremental Development process which uses the Sprint to represent its iterations. In other words, o e development cycle in Scrum is called a Sprint.

Every Agile methodology has a name for its iteration. So, Sprint is Scrum specific. Sprint is an iteration. However, an iteration in Extreme programming or Crystal or EED, etc, are not Sprints.

CHAPTER 5

WHAT IS SCRUM?

Getting familiar with Scrum

Looking to build a fast-moving, cross-functional team and engage every member that makes up your team work to achieve set goal, then Scrum is for you. Scrum is not designed to focus on individual member of a team.

Rather, it focuses on the team and how to harness their potentials to meet target. Like Jakes and his firm, making people get to work together to get things done in a more agile and efficient way is the way to go.

Scrum can simply be defined as the agile and efficient way to execute and manage a project, especially software development. As one of the process enhancement strategies in agile (we shall discuss agile shortly) and development rather than a methodology, Scrum is better viewed as a framework for managing processes. Scrum is an invention, a framework that helps team synergizes and works together.

Scrum refers to an iterative and incremental framework useful for managing product development. Scrum defines a flexible, full product development approach where a development team works closely and collectively to achieve a common target.

The framework allows and enables teams to self-organize by encouraging physical co-location or synergize knitted online collaboration of all team members, as well as daily face-to-face communication among all team members involved.

This brilliant, discursive, thought-provoking book is setting the tune for leadership and management process across various industries. The invention is practically changing not only the way we relate as co-workers, but also influencing our way of life and thought process.

It was clear from our pathetic story that the focus for Jakes and his team, over the years, should not have been on firing, hiring and headhunting. Energy should have been channeled towards making the management and top hierarchy bring together the various expertise and skills to bear on the company.

Scrum versus Agile: Differences and Similarities

One term closely associated with Scrum is Agile. We have discussed this extensively in our first chapter, but it is important we compare and contrast it with Scrum, one of its 'offspring's'. From the word go, you understand agile to be the umbrella term that consists of all methods and approaches that bring about change in the Development Team' process.

Let us see the relationship between Scrum and agile in this way: your kitchen's Dishwasher got spoilt by the high energy supply and you need to change it. You go to an electronic store where you see various models and brands of TVs.

There you see Samsung, Panasonic, Frigidaire, LG, Hisense, Tecno, Bosch and so on. Eventually, you settle for and leave the store with say a Bosch, perhaps because you think it has strong resistance to high voltage. That's exactly the same way Scrum relates to agile. Consider agile as the dishwasher, while Scrum is your preferred Bosch, one of the dishwasher brands.

But here is the contrast: while you may not be able to customize your dishwasher, agile processes, like Scrum, can be made to suit your needs. In that case, you can integrate other desirable features you find in other agile processes into Scrum.

For instance, you could employ components of Extreme programming, let's say test-driven development and pair programming into Scrum processes. That is the sort of flexibility and personalization feature that comes with agile. The flexibility of the agile process is one of those features that appeal to its numerous users.

Scrum is one of the many agile processes which also feature others like Extreme Programming, Adaptive System Development, DSDM, Feature Driven Development, Kanban, Crystal and more.

Agile means change, and it represents an umbrella name for all activities and approaches in software development. It is a term used to describe a general methodology used to achieve software development. Each of the agile methods, including Scrum, focuses on:

1. Teamwork
2. Frequent deliveries of working software
3. Close customer collaboration
4. Ability to respond quickly to change.

One of the key selling point of the agile processes is that it cuts a larger software project into several manageable breakdowns. Hence, it allows for the development team to handle projects in increments and iterations.

This is the point that separates the Agile management approach from other management systems. Agile like Scrum management, uses iterations at every phase of software development.

Agile process work with the proven results from studies that larger chunk of work does not most time yield expected result. Besides, studies have shown that the shorter a project, the higher its success rate.

The agile approach is to reduce as much as possible the size of the project in order to fashion out as many several smaller projects from it as possible. This will help the teams to manage and finish iteration easily.

Benefits of Scrum method

While it may be difficult to switch from say Crystal to DSDM or any other agile approaches, it is so easy using Scrum. Adopting the Scrum process has some positive feedbacks that apparently set it ahead of other agile processes. The benefits of using the Scrum frameworko are related and they build into each other. They include:

- Higher productivity
- Higher quality
- Reduced time-to-market

- Improved stakeholder satisfaction
- Increased job satisfaction
- More engaged employees

CHAPTER 6

ANATOMY OF THE SCRUM FRAMEWORK

What do you have to look out for as Scrum features?

The Scrum framework consists of a number of components that every leader needs to pay close attention to. Each part is integral to the overall functionality of the system to your company growth and process improvement. Besides, each module is integrated into the framework to serve a specific purpose and functions for the overall applicability and usage of the framework.

The Scrum framework is a merger of what is called Scrum Teams. The Teams have as their members associated roles, events, artifacts, and rules. Integrally, the rules are the bridges that connects the roles, events, and artifacts together. The rules serve as the guides, while if violated may lead to disruption of process and patchy operation and usage of the entire system and impact of constituent members. We shall now discuss each of these components.

Scrum Rules

Here are rules save you time, money and resources. The Scrum framework consists of basic frequently mentioned rules that guide the usage of the framework. As earlier noted, the rules are bridges that bind all other constituent members of the whole framework together. The rules clutch the Scrum process together so that everyone knows how to play and adhere.

Take the foundation away, the entire building collapses; remove the bridge, interconnectivity and communications break and chaos takes over. If you understand the meaning and purpose of chain action, then you appreciate how crucial the roles of the rules in the entire Scrum schematics.

Here is the great piece of news that may interest leaders who want to adopt the Scrum framework. And there is no reason why management should not adopt it. Interestingly, The Scrum system has the Scrum Master which helps in ensuring that everyone in the team follows the rules of Scrum as they relate to a specific project.

The main purpose of the Scrum rules is to achieve efficiency. Rules are set in the framework to ensure process improvement, optimize development systems, minimize waste, and effectively make use of scarce time and resources. Below is a list of basic rules in the Scrum model:

1. Every Sprint is Four Weeks or Less in Duration
2. There are no Breaks Between Sprints
3. Every Sprint is the Same Length
4. The Intention of Every Sprint is "Potentially Shippable" Software
5. Every Sprint includes Sprint Planning
6. The Sprint Planning Meeting is Time boxed to 2 Hours / Week of Sprint Length
7. The Daily Scrum occurs every day at the same time of day
8. The Daily Scrum is time boxed to 15 minutes

9. Every Sprint includes Sprint review for stakeholder feedback on the product

10. Every Sprint includes Sprint Retrospective for the team to inspect and adapt

11. Review and Retrospective meetings are time boxed in total to 2 hours / week of Sprint length

12. There is no break between Sprint Review and Retrospective meetings

Scrum Rules versus Generally Accepted Scrum Practice (Non-Core Scrum Rules)

Note must be taken not to confuse the Scrum rules with what could simply be regarded as the Generally Accepted Scrum Practices, abbreviated as GASP. It could also be regarded as the Non-Core Scrum Rules (NCSR).

While a Scrum rule is an inviolable process that if a team fails to do, they aren't doing Scrum, a GASP is some activity that sufficient number of Scrum teams are doing. We shall present that shortly in a table form. That would guide us when using the framework.

But we shall first and foremost define, in a formal style, what we mean by GASP or NCSR. Here's a working definition we can use. Besides, conducting a review meeting of a sprint at the end of the sprint is a GASP or NCSR and not a Scrum rule. Yet, a tea would still be considered to be doing scrum even if they overlook a sprint review meeting. There are teams who enjoy as common practice an after-sprint mini review rather than involving in a serious and bigger sprint review.

For instance, for a software development team, members may choose to conduct a short review in milestones with their product owner after each iteration is complete. After the review, the team may then discharge the new functionality to the website immediately.

Alternatively, the team may at the end of the entire work choose to have a review of the project in which case, they carry out all sprint review once at a time at the end of the entire iteration. Whichever method the team applies, it stills involves in Scrum.

What is GASP or NCSR?

Generally Accepted Scrum Practices (GASP) or Non-Core Scrum Rules (NCSR) are a set of activity carried out by a large number, and not necessarily all Scrum teams. Meanwhile, a team is not performing any activity would still be considered to be executing Scrum.

Teams' activities in the form of short, time-boxed iterations rather than a calendar month are not considered as GASP. Rather, it is a Scrum rule because, usually, Scrum encourages time-boxing rather than hourly estimation.

So, for any activity to pass as a GASP, it must be something that is generally accepted as a good idea. That goes to define a generally Accepted practice to mean that "every Scrum team should be aware of not involved in the practice.

Meanwhile, their non-involvement does not mean they are not excluded. Instead, they could elect to engage in some other practices outside what other Scrum teams do.

CHAPTER 7

USER STORIES AND CADENCE

A user story refers to a concise, brief descriptions of a feature. It is often told by and from the perspective of the user or customer of the system. It also means a short statement of a product requirement or a customer business case told by a product owner.

A user story can be written either by the end user or the team including the product owner, development team of Scrum Master. It should be expressed in plain language to help reader understand the power of the software.

The person who desires the new capability often writes a form of review of feedback as to how a system works. End users are often encouraged to discuss systems that they use rather than stating their features. User stories are often written on index cards or sticky notes and are saved in shoe box. They are also organized on walls or tables in order to facilitate discussion plans.

Telling a user story often follows a pattern like:

"As a user (you can mention the category of user you fall into)/ I want/can (state the goal of your desired product)/so that (give reasons for waning the product)."

Example of User Stories:

- As a driver, I want to be able to accept credit card payment so riders can pay seamlessly.

- As a rider, I want to add my Master Card to my profile so I can pay without cash and

- As a driver, I want to be able to upload my profile photo and that of my vehicle so that I can attract more riders.

- As a rider, I want to be able to view as many as possible vehicles so that I can chose from a pool one that is suitable for me.

Who writes user stories?

This is one important question that has one single straight forward answer: anyone can write user stories. The first step is that the product Owner makes sure that that he creates a Product Backlog of Agile User Stories.

However, writing the User Story proper requires all hand to be on deck. It has been assumed in some quarters that the end user is always the one who writes a User Story. This is a wrong assumption. Any member of the team including the Product Owner, Team, Scrum Master, or external members like the end users of the stakeholders can write them.

Each team member in any Agile project as a matter of necessity is expected to have user story examples written by them. The

Figure 1: Examples of User Stories for Websites

conversation that generates stories is more important that who writes the story.

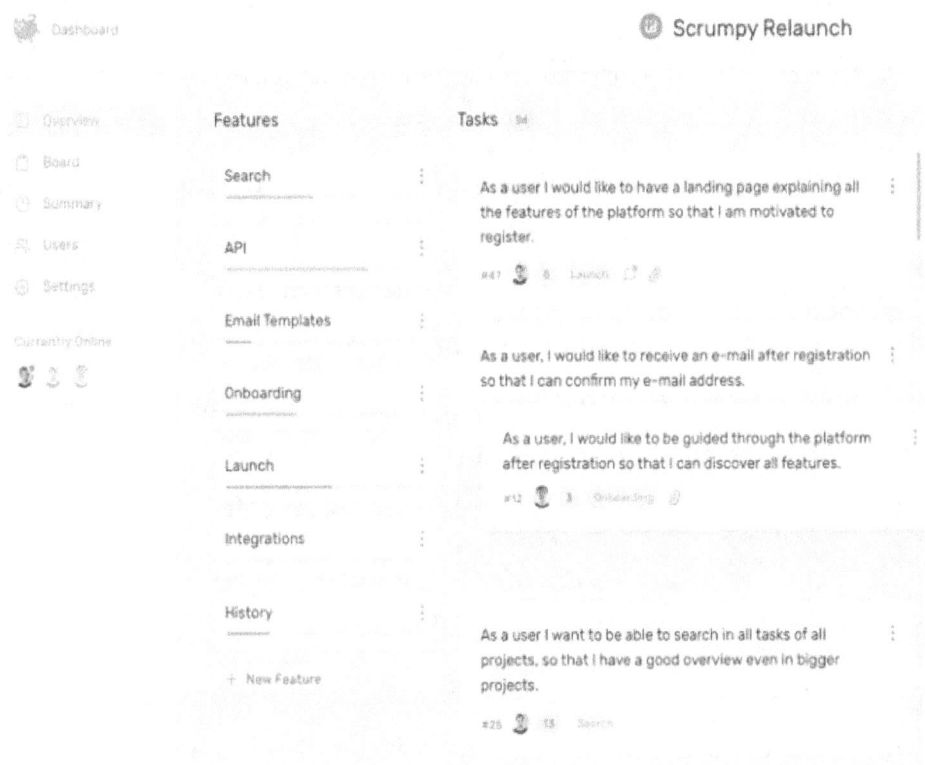

Tips to write excellent User Stories

You don't just write or tell a User Story. Consider it as an important aspect of your team project. In addition to the fact that you need to format the user story in such a way that it is always written from the customer's point of view, you need to take some important guidelines that will help you create a perfect User

Story. This will save you a lot of time discussing what exactly has to be built.

Here are a few trusted guidelines when generating a User Story:

1. Tell a story from Users Side

As the name suggests, a user story gives a description on how a product impacts on a customer. Either told by the user or the team, a user story must be written from the perspective of the software user or consumer. In another way, a user story describes how a customer uses or employs a product. Either way, the product must make impact on the bottom line of the user, and the story must be told from their perspective. Additionally, too, user stories must be targeted to capture a specific functionality of a product.

2. Tell a story; don't give a task

A story is not the same as a task. So, when telling or writing a user story, it must express feelings and give feedback. However, a user story requires several tasks to be able to deliver desired result and make needed impact. Here is the thing: task is concerned about implementation; user story is about definition. User stories must tell the 'what' about a product, not 'the how.'

3. Be simple, but stay high-level

Staying 'high-level' dos not suggest that your user story be isolated from your audience. No. You need to be accurate and straight to the point. The user stories must be written in plan, simple and solid language.

Doing this would help the team and stakeholders appreciate the needs and expectation of the user. Hence, it will help them avoid unnecessarily spending time explaining difficult terminology, buzzwords, and abbreviations.

4. Understand the users

It is important to stay up to date with trends and consumer needs. This will help you find out and study who the real users of your product are. From there, you can capture their expectations, get their profiles and points of view, and discover the associated 'pain points' using the software. Researching on user needs and other techniques could help generate insight into a better understanding of the key and real users.

5. Use epic stories

When writing user stories, one thing that is important is the use of epics. Epic user stories make a lot of deep impression on the reader. There are a number of ways you can use epics to convey your stories. One way is to take from across various different sprints, large amount of work to describe large pieces of functionality

Alternatively, since epics are for organizing stores and providing a bigger picture of impact, epics can come as grouping related smaller stories together to serve a common goal.

6. Prioritize stories, don't discard them

One thing to do to have great user stories and user story impact is to keep enriching your product backlog with new user stories. Keep at describing new interaction scenarios with user, get at random ideas and include out of product impact activities.

Effective prioritization process also involves a proper grouping of new entries. As great practice, it helps to manage the potential noise. So, do not filter out or discard items from your backlog.

7. Prepare for success not just acceptance

One area many scrum teams don't consider is how a user story should affect acceptance of product. Beyond the 'it works' or 'oh, software X is a great product' feedback, you often get from users, your user story should help generate metrics which show direct user feedback and capture how your product makes users happy and engaged. Yes, acceptance helps you gain a good control of the development life-cycle of the feature; success makes a mid and long-term impact and value on the real users of your product.

8. #Tag stories

Depending on the complexity of your product, you need to use as many as hundreds of user stories. And if you use multiple user stories, it becomes easier for people to navigate and get along with hem if only you use hashtags.

Doing this will require you to name, organize, categorize and tag your stories. One mistake you should avoid is to rename or change a story description after the initial few revisions of a story. You'll be bringing confusion to the audience and gaps in the team because of the inconsistency it will create.

Properly manage the metadata of your stories—status, progress, links, priorities, resources etc. This will help you explore, monitor, and understand your backlog.

Processes of a User Story

Typically, a User Story must go through three different important stages before it can be considered accepted. None of the three steps must not be missing in any User story you write. These features are Card, Confirmation and Conversation.

Card

The Card, otherwise called written text of the User Story, is a form of invitation and an upfront notification for conversation. One understanding about scrum is that the team does not necessarily have to write out all the items in the Product Backlog perfectly all at once. Changes and modification would occur along the line. Hence, the Card is an acknowledgement that both the customer and the team will continue to discover new areas of business needs as they are working on it.

> As a <user role> of the product,
>
> I can <action>
>
> So that <benefit>.

Conversation

The conversation refers to the collaborative discussion which is always facilitated at the expense and call of the Product Owner. The conversation involves all stakeholders and the team. The conversation can either be verbal or documented.

The conversation covers a wide range of issues. But most importantly, it is the hub of the real value of the user story. Issues raised during the conversation generate some of the things to be included in the user story. Hence, at every point in time, the Card should be modified to reflect the outcome and shared view of stakeholders and the team of this conversation.

Confirmation

The role of the Product Owner becomes more important here. No user story is consider done unless it has been confirmed by the product Owner. The team and the Product Owner check whether a story is complete or suitable for the purpose which it is set to achieve.

This assessment and eventual confirmation should be done in line with the Team's current definition of "done." Meanwhile, in the event that existing acceptance criteria do not consider the current definition of "done" in the eyes of the Team and Product Owner, new criteria should be established and inculcated to meet individual stories.

However, existing criteria must be well understood and agreed to by the Team before they are approved.

The INVEST criterion of User Story

Summarily, every User Story should fulfill the INVEST criteria proposed by Bill Wake, INVEST. INVEST stands for

Independent, negotiable, Valuable, Estimable, Small, and Testable, and each is explained below:

Independent —User Stories are actually the smallest piece of work that can be told in any sequence. It means a change to one User Story doesn't affect the others. Each User Story is a unique self-sustaining piece of work.

Negotiable – There is no rigid or fixed workflow on how to execute User Stories. It is up to the team to unanimously agree how to carry them out.

Valuable – Each User Story represents some value to an end user and so must deliver a detached unit of it to them.

Estimable –The team can seamlessly guess the amount of time it will take to complete the development of a User Story.

Small – Each User Story must as small as they are must go through the whole sprint cycle of designing, coding, and testing.

Testable –The Team should establish a set of criteria that for acceptance to confirm whether or not a User Story is implemented appropriately and accepted widely by user.

How to split Agile User Stories

Interestingly, you don't have to be creating new stories all the time. Sometimes, you may need to spill a user story into the next line, especially if a user story is too large to fit into a Sprint. The best and simplest approach is to split it so it looks like an implied conjunction. Use words such as "and" or "or" in the text of the story to create two or more new stories from the parts.

There are various ways you can split a story, and these include: splitting by:

- Process step, that is, taking each step as a new story
- I/O channel – making each I/O channel a separate story
- User options – making the options become user stories
- Data range – that is, every range, whether by year, month or digit, becomes a new user story.
- CRUD action – create, read, update and delete. This is applicable only if action is related to business logic.
- Role/persona: each role becomes a separate story.

Cadence

Cadence in Agile is defined as the number of days or weeks that is contained in a single sprint or release. Put differently, cadence refers to the length of the team's development cycle. In recent, the length of time to make a complete sprint cycle has changed from organization to organization. The business environment has become pluralized that companies can decide how many days make a cadence.

Ideally, most organizations use a two-week sprint cadence. The cadence that a project or organization selects is always informed by a number of factors including risk, project type, and how vital, decisive ad crucial the project is.

The table below shows an example of cadence:

September 2019

Sunday	Monday	Tuesday	Wednesday	Thursday	Friday	Saturday
			1	2	3	4
5	6	7	8	9	10	11
12	13	14	15	16	17	18

CHAPTER 8

SCRUM FLOW

The science of the scrum project implies that it follows a procedure. The whole systemic flow begins with a project vision. The vision defines every other thing that is to be developed. Certainly, the vision may look overcast from the start but as events unfold, it begins to take shape and becomes clearer.

The vision can be redefined over and over again, from market-based terms to system-based terms. The flow also consists in role assignment, with the Product Owner shouldering the responsibility of setting and delivering the vision to the project stakeholders and financiers with a view to maximizing their return on investment (ROI).

Furthermore, the Product Owner makes sure that he formulates a strategic plan for following up on prioritized project in the Product Backlog. Since the Product Backlog consists of a list of functional and nonfunctional requirements that deliver the project vision.

In the scum flow, the starting point is the organization of the Product Backlog. The Product Backlog is organized in such a way that items that will add value are given top priority with clear proposed releases. Organizing the Product Backlog consists of contents listing, prioritizing value-added items, and organizing of the Product Backlog, and then the proposed releases.

The Product Owner can effect changes to Product Backlog based on business requirements. The changes also depend on the speed level of the Team's ability to transform the Product Backlog to functionality.

All works on the scrum flow is done in sprint. Sprints refer to an iteration consisting of 30 repeated calendar days. Sprint consists of Sprint Planning, Sprint Review, Daily Scrum and Sprint Retrospective (See chapter 12 to read about the Scrum ceremonies).

There is a mutual collaboration between the Product Owner and the Development Team in the area of selecting priority items from the Product Backlog. While the Product Owner informs the Team what items in the Product Backlog he desires to be done; the Team in turn clarifies to the Product Owner how much of Product Backlog item desired list can be achieved and turn into functionality.

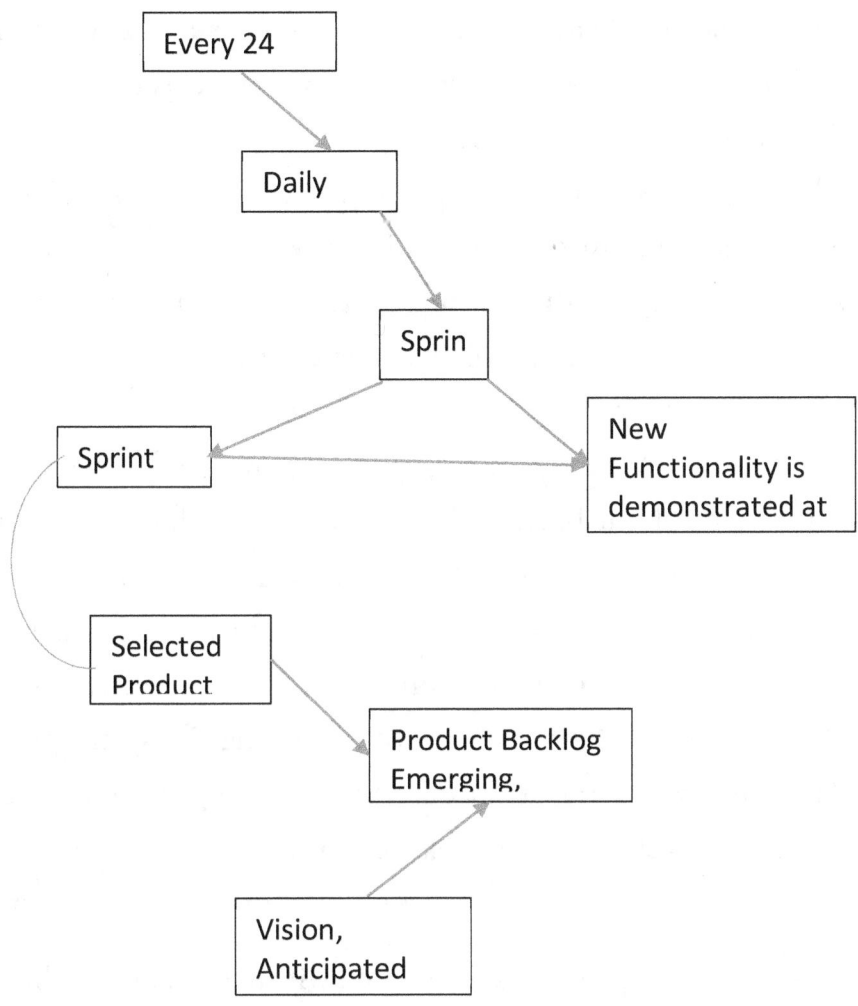

Scrum Process Flow

In our previous chapter discussion, we highlighted Scrum ceremonies, that is, the different activities and event that a scrum team undertakes before the completion of a sprint. How do these ceremonies relate and interact? Which one starts and which one ends the process? Now, we shall be having discussions around the flow of events during a sprint.

Simply put, Scrum process flow refers to the way to execute Scrum in the most effective, efficient way possible. It means the step-by-step procedure whereby teams can use the scrum to achieve efficient, high-quality results that optimize risk and maximize product delivery.

The scrum process flow starts with creating a product backlog up until we reach the backlog refinement stage, which entails review and revision of items in order to add detail, estimates and order. During the scrum process, too, the product development team carries out the refinement details while the product owner update the backlog refinement at any time. During the sprint planning, the point where the team determines the sprint goal, the plan usually answers questions relating to

- What items can be delivered within the sprint increment?

- How will the work completed within this sprint achieve the increment goal?

- How will the chosen work get done?

It is at this stage to that the development team decides what task is to be completed within a sprint and how it should work, informing the product owner and Scrum master the product backlog priority items.

There is the daily Scrum, a meeting conducted by the development team and intended to enhance communication, remove endless meetings, identify bottlenecks that could lead to project impediments, and then promote quick decision-making. The daily scrum allows the team to follow up on activities completed since the previous meeting. It also provides platform for team to prioritize work in the product backlog. The relevant questions to be answered at the daily Scrum meeting include:

- What did I do yesterday to help the team meet our goal?

- What will I do today to help the team meet our goal?

- Are there any bottlenecks or roadblocks that prevent me (or the team) from meeting the goal?

Scrum tasks

Scrum tasks are the subtle bits of work that are required to be done to give a finish to a story. Usually, a scrum task might taking the scrum team four to five hours to complete

On scrum task, team members can undertake to carry out tasks according to their skills and expertise. The sprint indicates the time remaining on a task and, on a daily basis, members who have offered can track the hours left to complete the task on the sprint. However, if members undertaking the task are unable to meet deadline, then the remaining task is riven into additional tasks. Meanwhile, until the tasks is complete, a story is not added. It is (complete) task that informs the addition of story.

How to create and add Task

Usually, a task is added to an existing story using the Story form. There are specific locations on the Form where scrum tasks can be added. They are, the Tasks related list and the Add Scrum Tasks related link. However, scrum task can also be added using the planning board and the story progress board.

Creating Scrum Task

To create a scrum task using the Add scrum Task location on the Story Form, follow the following steps:

1. Navigate to Agile

2. Then> Stories > Open Stories

3. Open the desired story.

4. Click the Add Scrum Tasks related link.

5. Set the number of tasks to be created or added in the dialog box that appears:

    o Analysis

    o Coding

    o Documentation

    o Testing

6. Click OK to create a batch of tasks of the selected types in the Scrum Tasks related list.

Scrum tasks created with this method are not yet complete and must be updated to become functional.

Open each scrum task record with a short description of ToDo and define the task.

7. Complete the form as described in the field description table.

8. Click Update to save your changes.

Eight Steps to a complete Scrum Process Flow

1. Determine the Product Backlog by listing product items and requirements in order of priority. This stage is carried out by the Product Owner in conjunction with Scrum Development Team.

2. Make estimate and plan for the workload based on the product Backlog items during the Product Backlog Refinement Meeting. This is carried out by Scrum team Development.

3. Hold a Sprint Planning, a meeting intended to define the sprint goal of the current iteration. Iteration is a duration of a Sprint, typically from 1 to 4 weeks. Select a list of User Stories to form the Sprint Backlog for the next sprint which could help to achieve the sprint goal.

4. Complete the Sprint Backlog, giving each member of the Scrum team unit tasks based on the Sprint Backlog.

5. Conduct a Daily Scrum, which is a meeting required to discuss progress and make review. Each of the Daily

Scrum is time-boxed, usually within 15 minutes. Every member of the team must speak face-to-face to discuss and interact with other members. The Daily Scrum always focus on past (what the team did and achieved yesterday), present (what it wants to do and accomplish today) and roadblocks (what could impede achieving Sprint goal), and update (reviewing team Sprint burn down chart).***[1]

6. Organize for each day a daily scrum that can be integrated and successfully compiled and showcased. Only release the version if the team endorses all of them and the unit test code is executed immediately.

7. The completion of user stories means the completion of the Sprint Backlog, and that signifies the end of a Sprint. After the completion of a Sprint, there is need to conduct a Sprint Review Meeting in which Product Owner and the customer must participate.

8. Finally, the Sprint Retrospective will be held after the sprint review at the end of each sprint. During the Sprint Retrospective, the team identifies by itself elements of the process that did or did not work during the sprint, along with potential solutions.

_____

CHAPTER 9

SCRUM BURN DOWN AND VELOCITY

The Scrum Burn down Chart refers to a visual estimation tool which indicates the work done and completed in a day against the projected rate of completion for the current project release. The purpose of the scrum burn down chart is enable the team to track the progress of project ad deliver the expected result within the desired and stipulated schedule.

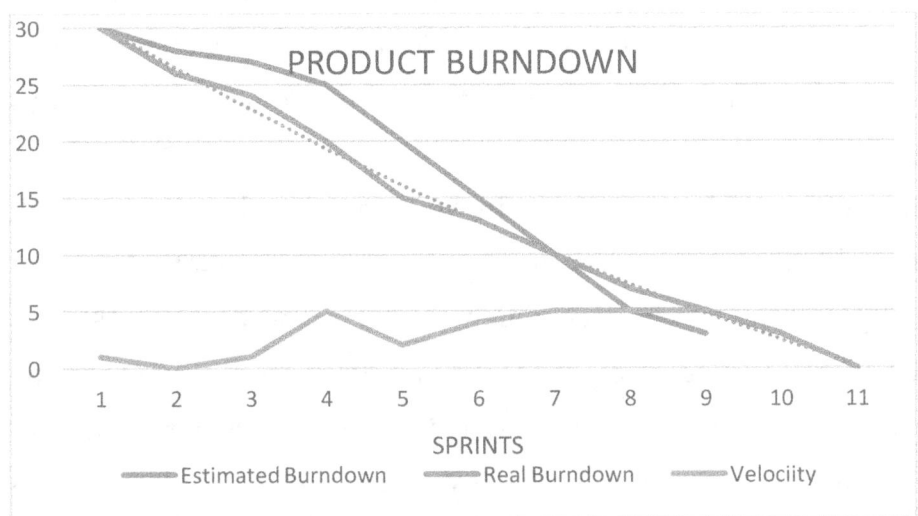

Figure 2: A sample of Burndown Chart

There are ways to execute the Burndown chart. It is important to know that stories in the Scrum should be burned down only in points and these have to be small. Secondly, planning poker must be used to estimate tasks in points and tasks must necessarily be burned down in points. Thirdly, the team must ensure that tasks burn down must be in hours.[2]

Velocity

The rate of progress of a Scrum development is called 'velocity.' It expresses the amount of story points or work a Scrum development can get completed per iteration or in a single sprint. In a more explicative form, velocity is an optional, but often used, indication of the average amount of Product Backlog that is turned into an Increment of product during a Sprint by a Scrum Team. Velocity is connected to the Scrum burn down because it is rate the Development Team within the Scrum Team works to achieve a project.

As events unfold, the entries in the Scrum Product Backlog will most likely change over time within the time allotted for the completion of the project. This would arise because new stories will be added; existing ones in the Product Backlog can be changed or deleted, depending on the Product Owner. Hence, in the simple Burndown Chart, the velocity of the Scrum Team and the change in the scope becomes indistinguishable (as we see in Figure 2 below).

---

[2] However, Scrum Foundation and ScrumIn no longer consider burning down tasks in hours as the best practice

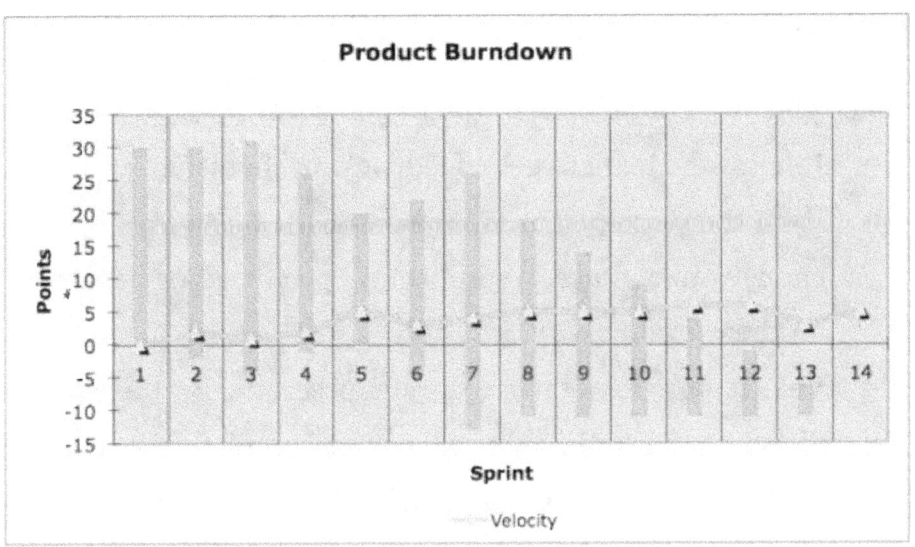

Velocity is a measure of the amount of work a Team can tackle during a single Sprint. Velocity is the key metric in Scrum. Velocity is always and should be tracked and calculated throughout the Sprint indicated on the Sprint Burn down Chart. The result of the calculated velocity should be visible to all members of the team. In that way, they will be able to measure whether or not changes they make are adding value to their productivity or not.

Ideally, Team velocity oscillate from one Sprint cycle to another and steadily trends upward by 10% on the Burn down chart in each Sprint. It takes the Team to complete three Sprints before it can determine its Velocity accurately. The Team and Product Owner must take their time to explain this to stakeholders, because the latter may not be ready to exercise patience until the three sprints elapse.

Purpose of the Velocity

Velocity serves a number of purposes including:

- It helps the team to get a feedback mechanism.
- With the velocity, the team can measure whether process changes they make are impacting positively or negatively on their productivity.
- Velocity also facilitates accurate forecasting by the team in terms of how many stories can be completed in a single Sprint. The Sprint forecasting is what is called *Yesterday's Weather* in Scrum.
- Velocity also helps in realizing Release planning.
- By knowing the Velocity, a Product Owner can be able to identify the number of Sprints it will take the Team to achieve a set level of functionality available for shipping.

How to calculate Velocity in a sprint

The velocity is calculated by counting only the user stories that are completed at the end of the iteration. It is forbidden to count the amount of work partially completed. An example of an incomplete work is to have a coding without testing.

Meanwhile, it is likely predictable to calculate the velocity of a Scrum Team and project the result after a conducting a few sprints. That is, Scrum Team has a latitude to give an estimated time left until all entries in the Scrum Product Backlog will be completed. However, it is not advisable to calculate points from a partially completed sprints. For stories. For instance, if the velocity of a Scrum Team is say, 30 story points and the total amount of work remaining is, say 155, it is easy for the team to put a figure of about 6 Sprint before it completes all stories in the Product Backlog.

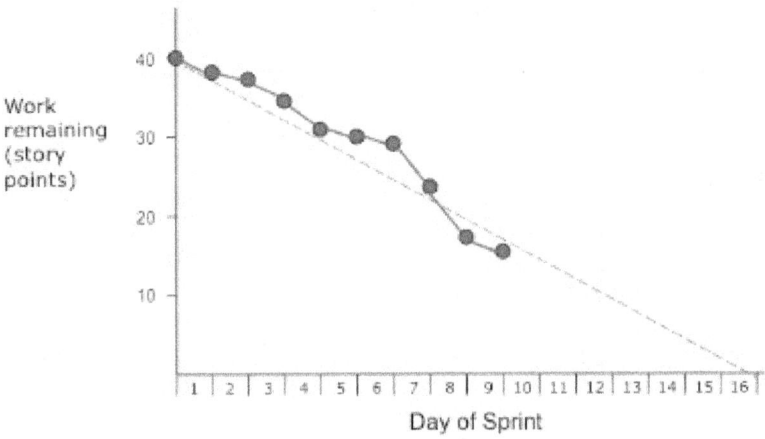

Day of Sprint

Question 1:

Calculate the velocity of the team in the following Burn down if, at the end of sprint, the team is only able to complete stories 1, 2 and 4.

Note: the team completed 50% of story 3, 40% of story 5.

Story 1 has 10 points

Story 2 has 4 points

Story 3 has 7 points

Story 4 has 3 points

Story 5 has 8 points

Solution 1:

The first rule in a burn down a partially completed story does not count. The team can only measure a potentially usable product increment. Hence, the velocity for that sprint would be 17.

However, if the team is able to complete the remaining 50% of the story 3 and 60% of story 5 in the next sprint, they would add 15 more points to make the velocity 32 points.

Question 2:

Calculate the velocity of a team in the following stories if the team is able to complete stories 1 and 2.

Story 1 has 5 points

Story 2 has 4 points

Solution 2:

The velocity for that sprint is 9.

Differences between the velocity chart and burn down chart

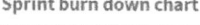

Sprint burn down chart

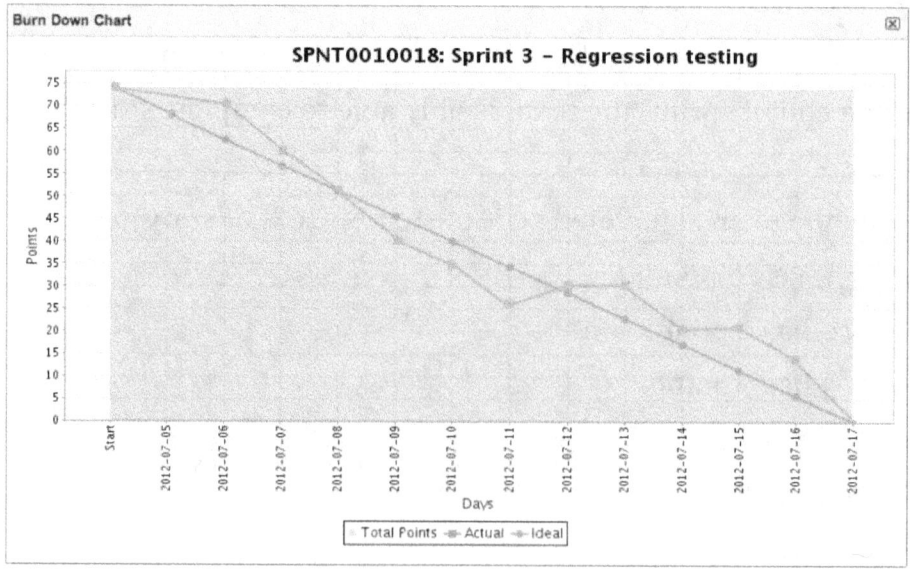

Figure 3b:

In the example in the chart above, the blue line indicates the ideal progress for the sprint from start to finish while the red line stands for the actual progress the scrum team made during the sprint.

Team progress just below the blue line is where team activity surpassed the ideal slope as team members completed more story points than desired. The upward slope shows the introduction of additional stories which means additional points. The team was able to complete 15 points of work in the final day of the sprint.

Burn down chart

The burn down chart shows the virtual progress a release team is making on a project in a sprint from start to finish against the real time actual daily progress.

The purpose of the burn down chart is to help the scrum master to be able to manage the releases and sprints in a more efficient way in a day-to-day fashion

It helps the scrum master to be able to track and address issues coming up.

Velocity chart

The velocity chart indicates the estimated effort calculated in story points that a release team is able to delivered across multiple sprints.

The chart provides the scrum master an insight into the general ability of the development team over time.

The velocity chart helps determine how many points worth of work can be completed per sprint for a given team

It allows for more accurate sprint planning

The Burn down chart displays comparisons of outstanding work against available time.

A team velocity chart shows the effort (as points) for a specific team against multiple sprints and multiple releases.

Velocity charts for releases display team performance across the sprints in a specific release.

Figure 2a represents Velocity chart while *Figure 2b* represents Burn down chart

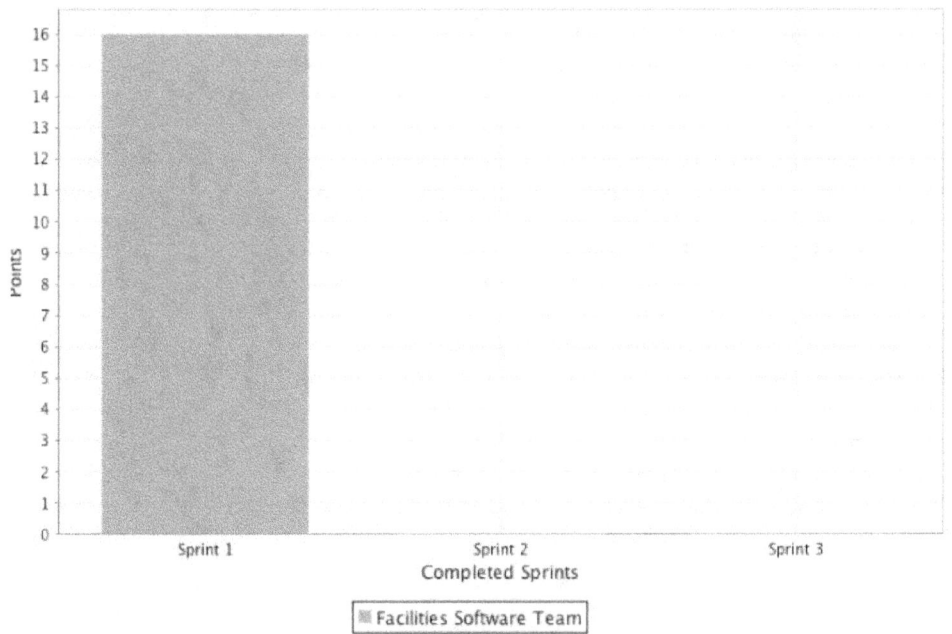

Figure 2a:

CHAPTER 10

SCRUM ARTIFACTS

This is one component part of the Scrum framework that you wouldn't want to miss. Let us go a bit archeological for a moment. The nomenclature may be the same, the role and meaning of artifacts in archeological studies do not also change as it applies to the Scrum framework. However, different system with different modes of application of terms.

What do artifacts mean in archeology? They refer to objects that are made by humans. Simply put, artifacts are products of human creativity. They are works of art, made either in the form of a tool to solve a problem or as an invention intended to inspire.

Similarly, the Scrum system is designed with some artifacts. Basically, there are three primary artifacts that the Scrum framework describes. They are:

- The Product Backlog
- The Sprint backlog
- The Product Increment

Other artifacts include Sprint Retrospectives and Product backlog Refinement. Each of these artifacts is for the umpteen time integral to the overall usage of the system.

The purpose of the artifacts in the Scrum is essentially to maximize transparency, and promote a share understanding of the work in a team. While the Sprint Backlog and Product Backlog are responsible for defining the work to be done that will enhance work value, the Product Increment refers to the achieved portion of the work done work during a particular duration or sprint.

The Product Backlog

The Product Backlog is the portion of the Scrum project that sets scale of reference. By that the segment answers the fundamental question relating to 'what is the most important task to build next? In simple terms, Product Backlog can be described as constantly evolving artifacts that are ongoing.

The segment serves as the roadmaps that set the tone for what is to be done at each stage of the project. It is important to note that the Product Backlog is open to modification, constant update and refinement in order to suit the changes that occur as a result of dynamic nature of product development.

Managing your product backlog

This is one thing many Scrum practitioners have to get familiar with. It is crucial to always make effort to keep your product backlog small and manageable. You're likely to face three key intractable issues if you keep stocking your product backlog with too many items. They include:

- **Consumes unnecessary time**

Working with too many items in the product backlog makes work delivery harder and in the process large amount of time is lost. This is because the time it takes to sort out the items packed up haphazardly in the backlog is always longer than the time it would take to arrange. Prioritization becomes harder and there is higher chances that items are duplicated.

- **Team Progress is hardly noticeable**

Due to time lost in packing up items in the product backlog, the Scrum Team hardly notices progress they make. For instance, a team that finishes 20 of 70 items are likely to see the progress they make. However, when a team completes 20 of 900 fee frustrated as the sense of accomplishment and drive to continue would obviously diminish.

- **Shortens human capital**

Imagine a five-member team having to carve out two members to concentrate sorting out items in the product backlog. That approach is unhealthy for speed, urgency, and achievement. If someone has to spend valuable time creating all those product backlog items, then it reduces the chance of achieving the visibility projected into the future from the start of the work.

In order to avoid facing the above snags, what must a team do? To be able to effectively manage your team's product backlog, keep it small and never have it item-clogged, the following steps will be useful.

Remove items you'll never realistically do

Once you find out that there are items in the product backlog that the team would not realistically be able to do, it is advisable to ruthlessly purge them as quickly as possible from the product backlog. In that way, you will be able to keep your product backlog small and manageable.

Although this may be hard to achieve, and sometimes come with some shock situations in which you need to make contingency, a leader should be highly proactive in thinking ahead of team members in generating on-the-spot ideas to rescue the situation.

Keep off 'not-at-the-moment' items

One other strategy akin to the above is keeping off the product backlog items that the team is not ready for now. Yes, the product owner wants those items, but is the team currently ready to work on them? How fairly soon would the items be needed? Is the owner ready for pay for them now?

If the answer to the first is no, then take them off the shelf. If it is yes to the third, then retain them and keep off other ones that are not needed right away. If the team is not disposed now to work on them and would not be ready for them in the nearest foreseeable future, kindly delete them off the product backlog.

Instead of clogging your product backlog, create a holding tank where you can keep standby items until they are ready to be treated by the team. Doing that would keep your product backlog small and manageable.

Product backlog needs periodic review

Consider the product backlog as your wardrobe where you keep the clothing. How often do you check? I guess daily. That is exactly how you should treat your product backlog. If not daily, as you do your wardrobe, at least make it a point of duty to periodically (maybe quarterly) review your product backlog.

Keeping your product backlog to a reasonable size is no brainer. Initiate a regular review process, to check the fancy and non-fancy items. In fact, the product owner can help in this regard, helping you to clean up, delete, or move items that the team won't work on or items that do not drive immediate attention of the team.

Product Backlog Refinement

Before we move to the next Scrum Artifact, it is important we discuss the product backlog refinement process. Our previous knowledge puts us in the understanding that Product Backlog refers to constantly evolving artifacts that are never complete.

If some artifacts are ongoing, something must be the chief initiator of such work. This is the role the Product Backlog Refinement plays. So, Product Backlog refinement refers to the activity executed to constantly evolve. It becomes clear that the Product Backlog Refinement occupies a central and constant place in the life of the Product Backlog, because it keeps evolving new product backlog.

Sprint Backlog

As Scrum artifacts, the Sprint Backlog refers to a list of tasks the Scrum team has identified and itemized to be completed during the Scrum sprint. Usually, during the sprint meeting when a Development team sets out its plan, the team often chooses some number of product backlog items, usually in the form of user stories. Hence, the team identifies the tasks necessary to give a finish touches to each user story.

Also, the spring backlog can be understood to contain two quick calls to action. One, think of it as the 'How' of the Sprint and the 'What' of the Sprint. By the 'What', we means task to be completed by the team. The 'How' stands for the way such tasks would be delivered.

The Sprint Backlog represents a highly evidently planned out, real-time picture of the work that the Development Team plans to accomplish during the Sprint. The sprint backlog is an item only excusive to the Development Team

Besides having the two important components of 'How' and 'What' of the Sprint, Sprint Backlog also consists of the Development Team's design for how the team would carry out and deliver the product Increment.

It all depends on you, you could maintain your sprint backlog using any of the available software products designed specifically for Scrum. The Scrum framework also allows you to represent your sprint backlog as a spreadsheet or deploy your defect tracking system. The table below is an example of how you can maintain your sprint backlog.

The following are the important contents of the Sprint backlog:

1. Tasks that have been decomposed from the user stories and accepted by the Team for the current iteration.
2. Story points or time estimations for individual tasks.
3. Product backlog Refinements of the "definition of work done" as it concerns a specific story or task.
4. Product Backlog Refinements to stories that don't compromise the Sprint Goal or require the Product Owner to call for an early closure of the Sprint.
5. In-sprint stories or tasks added by the Team to give teeth to the current Sprint Goal.

Again, it needs to be emphasized that the Team could at the Spring Planning stage sets out the Sprint Goal and stories accepted for the current Sprint. However, Team updates and modifies tasks and sometimes stories on the Sprint Backlog when they see the need to do so in other to support the Sprint Goal.

Ideally, updating the sprint backlog per day is ok. However, as soon as new information is available during sprint, members of team can update the sprint backlog. Other teams could choose to do their update during the daily scrum. Usually, the estimated work left undone on the schedule is calculated and put into graphical representation using the ScrumMaster, and it will look like the chart below:

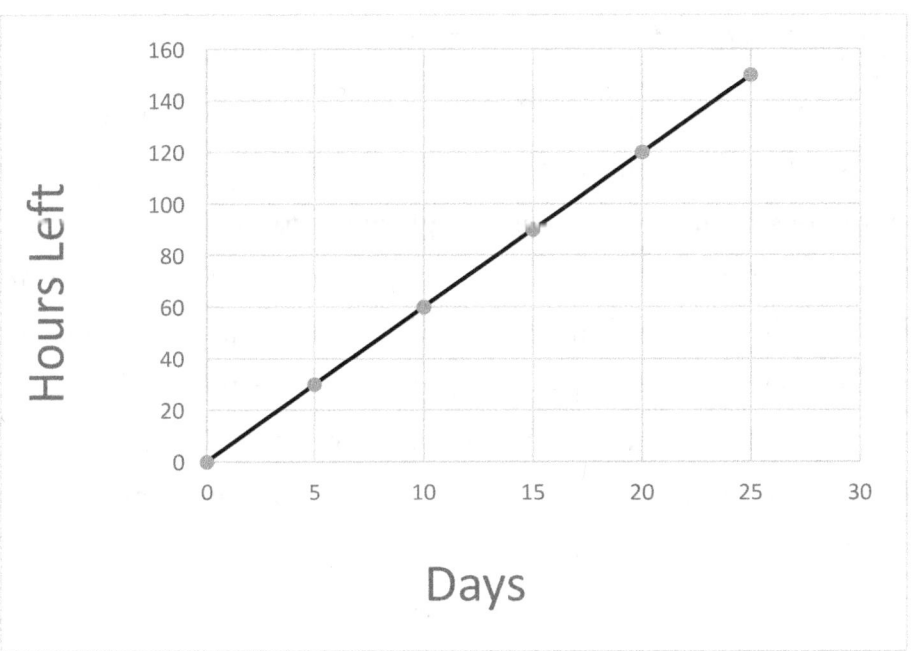

In the sprint backlog graph shown in chapter 1, what we observe is that the team in this scenario jerked in too much work at the initial stage into the sprint backlog. Yet, they had about sufficient hours to go on day 5 of a 25-day sprint.

Meanwhile, product owner had to add some user stories from the sprint, resulting in the big he rise on the chart from day 1 to 25. The graph makes consistent progress, finishing the Scrum sprint successfully

It is very crucial that the team pulls the accurate amount of work done, completed and remaining into the Scrum sprint. The team can also add or remove tasks.

Scrum Focus Areas

What do you have to focus on in Scrum framework?

- **Focus on Team Performance**

It is worth re-emphasizing that it is in team success you that find the skills and commitment of each member. After all, nobody commends perfect rehearsals; people applaud excellent output.

So, with Scrum, you're not looking for a strategy that sets to distinguish individual members of a team. Rather it seeks to identify and then harness the potentials, expertise, skills and experience that reside in individual members of your team. In doing that, Scrum focuses on team performance, measuring corporate and not individual performance.

- **Relative Task Comparison**

How do you measure performance of team tasks in your company? Estimating member's effort based on number of hours? That absolutely is not going to produce desired and desirable results. Scrum comes with an incredible different approach that works.

The system operates on the principle of relative estimation of task based on complexity and size. In other words, relative estimation preferably works better as a method of sizing up member's effort. The idea is that when you compare in relative terms individuals' tasks based on their complexity and size, rather than number of hours spent, the result you're going to get would be less prone to error.

- **Engage Time Boxing**

This is one novel approach project managers and expert company progress analysts find highly useful. Scrum focuses on gaining commitment and speed by time boxing. By that, the strategy evolves the practice of allocating task to a specific iteration.

Through this, each iteration has an equal, fixed duration within which each task is to be completed. This is in contrast to existing model—especially the Kanban, which achieves commitment and speed by limiting work in progress.

- **Evaluate Team Happiness Indicators**

This is a key feature in Scrum. Measuring happiness indicators of your team at the end of a project is a concept motivated by the idea that team performs better in future tasks once they express happiness and satisfaction with the outcome of the project at hand.

As a manager, therefore, it is important to tract, appreciate, measure, and evaluate the leading motivating factors that enhance the interest and happiness of the rest of the team.

- **Focus on goal not roadmap**

Scrum framework does not discourage the use of roadmap while setting out on a project. However, roadmap should rather serve as a guide and not a cast in stone kind of thing. With Scrum system, every sprint has its own uniqueness, and the roadmap can be modified to suit the goal of each sprint.

Put simply, roadmaps are loose visions at best. Historical performance and lessons have a crucial role to play to dictate or inform changes that would occur from sprint to sprint. This leads to what is called 'velocity' in Scrum system. By velocity, teams are believed to decide and define their capacity by looking at their performance in the past.

- **Adopt Commitment to Process Improvement Approach**

Commitment to continuous improvement of processes is a core aspect of Scrum. Jeff Sutherland's invention works with the goal that every stage of iteration ideally should produce enhanced results that improve on the outcome of last performance. Process improvement in Scrum is a strategy that focuses on milestone improvements, technically called *kaizen*.

CHAPTER 11

SCRUM CEREMONIES

Scrum Ceremonies

Scrum ceremonies remain one of the vital force and components of the Agile software delivery process. They refer to the gathering and coming together of Scrum team with a view to getting work done in an organized, structured manner.

As the name suggests, Scrum ceremonies are meetings not just for the sake of it of fun. Rather, scrum ceremonies offer the platform for scrum team to effectively empower, influence, brainstorm, collaborate, and help each other to grow and ultimately drive results.

If scrum ceremonies are not properly managed they can become mere jamborees where the team spend a lot of invaluable time gallivanting and overwhelmingly waste productive time of their calendars and whittle down the purpose and value of their coming together.

Another end for which the Scrum ceremonies are often organized is to realize and enable many of the Agile core values and principles, some of which include work progress review, customer satisfaction, sustainable effort, team work, effective communication, among others.

While it is important for team to hold these ceremonies regularly, sometimes, they may abandon them, either because the team does not see the need and value for them at the time or they have jettisoned the core values and principles. In either way, it is dangerous for team to not hold scrum ceremonies.

What are the Scrum Ceremonies?

Now, let us discharge the four scrum ceremonies and the purpose for which they are important to the overall scrum and agile drive. The key facilitating factors for effective scrum ceremonies include purpose, attendees, tips and tricks.

The best way to start this section is to proclaim unequivocally that Scrum is deliberately a simple and lightweight process. However, its mastering can become or be made difficult, especially if its cores goals of serving as a framework for cross-functional teams to unravel complex issues are not taken seriously.

Scrum ceremonies provide the platform where these complex problems can be discussed and resolved. They help expand the scope and bring niceties to unstructured framework.

It is important to bear in mind that these ceremonies are tied to specific activities and goals of the scrum framework. In simple terms, we could consider scrum ceremonies as a great Agile process that team deploys globally to develop strategy that is actionable. The scrum ceremonies include:

1. Sprint Planning
2. Daily Scrum
3. Sprint Review
4. Sprint Retrospective

1. Sprint Planning

Sprint Planning

Just as the name indicates, sprint planning consists of the set of **arrangement, scheduling and design** organized to make sure the development team is adequately and properly prepared to achieve a goal and get work done every sprint.

When the team meets at the beginning of a new sprint, the purpose is often to design for Product Owner and Development team to do a review of the prioritized product Backlog. One thing that is important to note here is that the prioritized product Backlog should contain only items that the team would be able to complete at the end of the sprint.

Sprint Planning as a Scrum ceremony, is essentially organized for the purpose of engaging in a series of productive discussions, negotiations, and brainstorming that ultimately target generating a sprint backlog that contains only all items the team is committed to achieving during the course of one single sprint cycle.

These achievable items could be tagged the sprint goal, a demonstrable increment of work that can be proven at the end of a sprint. The items should be agreed upon by all team members, in that way, they are able to work together towards achieving them.

Who should attend Sprint Planning Ceremony?

In simple answer, all scrum roles should be in attendance. Each one has an eminently important role in Sprint Planning facilitation in order to ensure a successful deliberation. As a matter of fact, every Sprint Planning has in attendance the Product Owner, Scrum Master and Development Team.

What does the Product Owner do during Sprint Planning?

Apart from the Scrum Master, the Product Owner shoulder a lot of pre-meeting responsibility, preparing the entire roles for the Sprint Planning. For instance, the Product Owner has the responsibility to prepare the Product Backlog and ready for review ahead of the meeting.

In addition, the Product Owner should add acceptance criteria, requirements and all other necessary details for the Development Team that would give a detailed and accurate estimate of the level of effort and performance.

All grey areas in terms of possible questions and assumptions that the Development team may have the product Owner should anticipate and clarify. In that way, they have a blueprint of what the team is going to do and achieve in a sprint.

What about Scrum Master?

Scrum Master is primarily responsible for facilitating the entire session of the Sprint Planning. They play the active role of ideal facilitator, making sure all questions and assumption of the Development Team which answer have been prepared by the Product Owner are adequately addressed. The timing, duration, Q &A session, and closing of the meeting are responsibilities due the Scrum Master.

And the Development Team?

The Development Team is not a spectator or onlooker during Sprint Planning. They form an integral part of the whole session. They must be reasonably critical of the outline and responses of the Product Owner. The prepare questions about items to be pulled into the Product Backlog for the entire sprint.

What is the length of Sprint Planning?

The length of a Sprint Planning ceremony depends largely on the length of the sprint. So, if the duration of your sprint is two weeks, then your Sprint Planning should not last more than 3 to 4 hours, max; for a week-long sprint, the Sprint Planning should last not more than 2 hours.

2. Daily Scrum

Alternatively named the daily standup, daily scrum refers to a quick pulse check that not only defines the work for the day for the team but more importantly illuminates the team to identify all roadblocks to team progress.

Also called a scrum meeting, this meeting offers the team opportunity to come together and define what the day's work outline should look like, identifying any impediments and prospect.

What's need for Daily Scrum?

The goal of the daily Scrum is primarily to do a review and progress overview of their daily activity. It is a platform that provides the team with a frequent chance to communicate individual and corporate progress. A

All of the talking and deliberation should be geared towards achieving that common goal set out at the beginning of the sprint. Daily Scrum helps the team to identify work blockers and proffer solution on how to remove them.

Who does what during Daily Scrum?

The Scrum Master has the responsibility of shouldering the clearing of bottlenecks to the achievement of goals by the Development Team. This would help the Development team achieve more and focus on delivering the work identified in Sprint Planning.

The Development team executes all planned work and activity. Although they take active part in the entire Sprint planning and Daily Scrum asking questions, giving possible roadmap, the Development Team caries out all activities post-daily scrum.

During each daily scrum, the Development Team gives answers to the following important questions:

- What did you do yesterday?
- What will you do today?
- Are there any impediments in the way?

Although the presence of the Product Owner is optional during daily scrum, he is strategic in preparing the entire roles ahead of the daily standup. They prepare answers to any blockers or roadblocks that the team may identify.

How long should a daily scrum take?

Ideally, since a daily scrum is to review, outline day's work and identify impediments, it should not last more than 15 to 20 minutes. However, it can stretch more than that depending on workload in the previous day. But, a daily scrum should be kept short and simple to allow the Development Team and other stakeholders get on to work early.

3. Sprint Review

In simple terms, sprint review means 'stakeholder, kindly have an assessment of our completed work.' That sounds too simplistic, right? That is exactly what it is. During Sprint Review ceremonies, all finished work are displayed by the Development Team to the stakeholders. However, the Scrum Master and Product Owner are not excluded in the session.

They come together to showcase the outcome of the work they have done during a sprint cycle to stakeholders to have their own assessment.

Why Sprint Review?

Just like you have to show your teachers at the end of the term how much study your have assimilated by going to write exam, so also at the end of each sprint, you have to be assessed.

Again, just to remind you of the routine in your old school days: as you're about rounding off an academic term, there used to be a window of opportunity where you have review of all that transpired. The same rule and procedure apply here.

The platform also provides opportunity for stakeholders to take a look at what has been done and have a sooner than later feel of possibly what to expect. The stakeholders can quickly adapt the product that form the result of the effort of the Development team.

It is important to keep in mind that work showcased during Sprint Review should be shippable to the extent that they meet the definition and scope of what was defined at the begging of the sprint. This would boost the stakeholders' confidence in the team and particularly the Development Team gets much of the thumb-up.

Again, Sprint Review is not time for the team to panic or shiver like an average PhD student preparing for his Thesis defense. No. Sprint Review, which is alternatively called Sprint Demo, is an interactive session where Development Team displays what they have been able to achieve over the course of a sprint cycle to stakeholders.

It is geared towards building trust and confidence the stakeholders repose in the team and thus strengthens the relationship between the two parties. It serves as the face-to-face manner for stakeholders to have an early feedback from customers and objective assessment of what has been done by the team.

Again, it should be done in a relaxed mood as it is intended to show the business value the completed work would bring to the product development. The team should ensure they do everything possible to impress the outsider reviewers and external evaluators

When is Sprint Review Appropriate?

The team can decide to have the Sprint Reviews staged on a causal "Demo Friday" or make it a kind of organized event where everyone is seated, looking serious!

Who is needed in attendance?

Everyone! Anyone! From the Product Owner to Scrum Master, to Development, attendance is crucial. This is the time the entire team is showing they are up to the task they are charged with. Their delivery would inform stakeholders how competent or otherwise the entire team is. Also in attendance are a blend of management, internal and external stakeholders, end users, and developers from other projects.

It is the responsibility of the Scrum Master and Product Owner to be engage in discussions on who should be in attendance during the Sprint Review. Interestingly, the Sprint Review is more open to attendees than other Scrum ceremonies. It is a fluid event that offers opportunity for insider and outsider assessment of the team.

How long should it take?

If you think of the duration for the Sprint Planning, then you get a sense of how long the sprint review should take. Preferably, an hour a week of the sprint should be enough. If for, instance, you have a two-week sprint, a two-hour Sprint Review should be scheduled.

4. Sprint Retrospective

If you're looking for a platform to get technical during Scrum ceremonies, then Sprint Retrospective is not the right place for you. Maybe, you can check Scrum Planning. However, Sprint Retrospective represents the final lap in the series of scrum ceremonies in which the team can look back do a thorough appraisal.

While other scrum events and ceremonies may offer you the not-too-serious, not-too-relaxed mood, Sprint Retrospective takes the anti-Aristotelian extreme virtue by giving us a too-serious platform. It is an opportunity for the team to ask questions about the feedback they get after showcasing the work they have completed. Here, the team looks back to see completed work and identify items that could be improved.

What's in for the Team?

Great question! After Sprint Review has been done and stakeholders and participants have given their respective review and feedback, the team needs to now sit down and see how they ca improve on subsequent deliveries.

It is during Sprint Retrospective that the scrum team can discuss situations that are going on fine; access the ones that need improvement and suggest possible ways to enhance work delivery. The fundamental questions that the team often confront itself with during the Sprint Retrospective include:

- What went well over the last sprint?
- What didn't go so well?
- What could we do differently to improve?

Sprint Retrospective Should Drive Change Not Blame

Sprint Retrospective should be an avenue not for blame game or castigating a team member for doing or not doing something; rather, the session should be a blameless space for members to give their honest and objective feedback and recommendations. Simply put, it should be a forward-projecting rather than a backward-looking space. Essentially it should drive desirable and desired change for the team. For this to happen, all recommendation, data and feedback got from members and participants should be collated and gradually implemented where appropriate to further strengthen the team's future performance.

Who should Retrospect?

At every Sprint Retrospective meeting, the Scrum Master and the Development Team. Should be in attendance. Meanwhile, participation of Product Owner is an optional thing. Outside stakeholders can be in attendance too.

What about duration?

Being a session for feedback, Sprint Retrospective meetings should be longer than a Spring Review in terms of duration. However, it should not exceed the maximum period of 1 hours thirty minutes, for a two-week sprint. But if you are having one-month sprint, then your Sprint Retrospective can take as much as 3 hours.

Below is a chart representing the maximum allotted duration for each Scrum ceremony facilitation for a four-week Sprint:

Summary of Scrum Ceremonies

Agile is focused on constant and gradual improvement and driving change, hence all ceremonies should be driven towards ensuring better quality service delivery. The team should be motivated at every session of the events to have a progressive approach to the work.

While the team is expected to improved, resources to be used should also be enhanced; approach should also become more efficient and effective to achieve more with little. Needless to say that Scrum ceremonies should be forums where every member is shown the capacity and potentials that reside in them.

CHAPTER 12

THE SCRUM CORE AND NON-CORE ROLES

There is nothing weird or out-of-this-world about the term Scrum Roles. It refers to the daily use of role assignment in our offices. But here is the difference, scrum roles are focused on the team and not on individuals.

They reflect the collective roles individuals play in achieving collective result. In scrum, there are basic groups who perform certain roles towards project management and software development. They include the product owner, scrum master, and development team.

Let us discuss each of the Scrum core roles.

1. The Product Owner

Allegorically, there is no building without a builder; so also, no project is executed in a vacuum. As the name suggests, the product owner is the person who owns the product. Maybe, that sounds circular. In Scrum, a product owner represents the client and the business for the product on which they're working. The product owner possesses the backlog. They determine which items take precedence over others. They are responsible for setting the tune for every sprint or iteration by striving to give priority beforehand to items to be worked on.

The survey the market and the industry to identify needs and expectations of customers and then make informed executive product decisions on a daily basis. In the process, they ultimately help in translating those needs into actionable work items for the Development team.

2. The Scrum Master

As much as a team desires that work must be delivered quicker and to taste, workers must be adequately equipped with all the resources they need to achieve this goal. So, there are workers; there are those responsible for making things work.

A Scrum master belongs to the second group. He is that person responsible for ensuring the team has everything they need to deliver value. A scrum master could be a coach, facilitator, motivator, moderator, mediator, counselor, advocate, or a mediator.

The Scrum master must be one who is capable of projecting into the future, foreseeing challenges before they turn into project impediments. Hence, he is an impediment remover.

He facilitates the process of making sure there constant and smooth communication channels among members of the team. As an advocate, the Scrum master serves as the middle man between development team and the product owner.

He mediates to ensure that all facility needed are in place before they are needed. As a matter of fact, a Scrum master coordinates all activities; he is the project manager in scrum. So, whatever duties and responsibilities you think a project manager should do, the scrum master should.

3. The Development Team

Delivering working software is no joke. It requires not just a vision; it must be driven by a team. A development team refers to a group of cross-functional and multi-purpose team members who are focused on ensuring delivery of working software.

The Development Team, also called the Scrum Team, consists of all persons involved in the technical aspect of a project. They include professionals, software development newbies, experts, novices, designers, QA who collaborates collaborate on the actual development of a product.

Usually, the development team is composed of 5-10 people who are fully dedicated to working out a scrum project. However, reality may change the course of things in which case agencies might take a different approach based on the challenges they grapple with.

In any case, ideally, the development team adequately aided by the facilitation of the Scrum Master and Product Owner, should be a self-organizing and self-motivated group of individuals who offers value.

Non-core Roles

Non-core roles are those roles which are not mandatory or required for the scrum project to run effectively. These roles consist of members who only interested and willing to be part of the project from an outsider point of view. They are not directly involved in the day-to-day running and core implementation of the scrum project. Unlike the core roles, the non-core roles have no formal role they play in relation to the scrum project.

They may assist and interface with the project team but do not have any official responsibility towards the success of the project. However, they non-core roles as we have seen in our previous discussions serve as the third-eye that keeps the team on its toes to be able to achieve optimal result. Hence, the non-core roles should be taken into consideration when the team is making decision on the Scrum project.

Their continuity may be ceased by the team at any time, but it is not advisable that the team scraps members of the non-core roles on any project. As we see in the Sprint Review session, the stakeholders are very important in giving feedback and ensuring the team goes back and improve on the project they come to showcase.

The non-core role participants in the scrum project include:

1. **Stakeholders**

By stakeholders, we refer to a number of interests. It is a collective term that refers to the assortment of individuals who volunteer to be part of the project. Some stakeholders participate in the project only as consumers; others take part as sponsors. Basically, stakeholders in the Scrum project include customers, users, and sponsors.

These people constantly interface with the Product Owner, Scrum Master and Development Team. They duty though no formal is to provide the core role members with valuable inputs that would help in improving showcased projects. They also facilitate creation of the project's product, service, or other result.

Stakeholders are also important person who could influence the project throughout its developmental stages. They play active role in other stages including Develop Epic(s), Create Prioritized Product Backlog, Conduct Release Planning, and Retrospect Sprint.

2. Customer

Every product has an end user and consumer. Product or service without patronage is dead on arrival. So, when a scrum project is being designed by the Product owner, there must be a consumer in mind who will purchase and use them. In fact, consumer has been described as the life wire of company sustenance. We may ask: who is the scrum project consumer?

The consumer is the individual or a corporate body that subscribes to acquire the product or service of manufacturer or service provider. In relation to scrum, the consumers are those persons and organizations who desire to use Scrum product.

Scrum products' consumers can be either internal customers, that is those within the team or external customers, that is, those who have no formal link with the organization where Scrum is being applied.

3. Users

There is a little difference between a consumer and a user. A user refers to the individual or the organization that directly utilizes Scrum project's product, service, and other result. At some time, a user can be the consumer; at other times, they are not.

However, like consumer, users form part of the non-core role participants in any organization. Users can be those within the organization or those outside. But whether an internal user or external user, the role of the user cannot be overemphasized.

4. Sponsor

In a way the sponsor is a major stakeholder in any project. The sponsor refers to an individual or corporate organization that is putting down the funds to keep a project going. They provide the resources, whether human or infrastructure as well as technical and operational support for the project. As a stakeholder then, the sponsor is one to whom the organization is also accountable.

5. Scrum Guidance Body

The Scrum Guidance Body (SGB) is one of the non-core optional role in the scrum project. The SGB is a collection of brochures and a group of professionals and experts who characteristically participate in defining objectives related to quality, regulations, security, and other key organizational parameters.

The objectives defined by the Scrum Guidance Body help to guide the work done by the team, which includes the Product Owner, Scrum Master, and the Scrum Development Team. Also, the SBG also supports capture the best practices that should be used during the implementation of all Scrum projects in the organization.

However, the Scrum Guidance Body does not have power to make decisions in connection with the Scrum project. The SGB only can act as a consulting body which only guides the hierarchy levels in the project organization

The structure of the Scrum Guidance Body serves as the advisory body for the portfolio, program, and project of the project organization. The Scrum Team may or may seek advice from the Scum Guidance Body. The optional role of the SGB does not mean that they are unimportant.

6. Vendors

Vendors are non-core role players in the project organization. They are internal or external individuals or organizations that provide products and services that are not available within the core competencies of the project organization. It is possible that vendors can also act as the same person or organization, playing the role of a stakeholder, sponsor, or customer.

CHAPTER 13

SCALING SCRUM

Disclaimer!!!

Here is the first thing to say, a sort of disclaimer before you even conceive the idea of scaling Scrum: Will scaling resolve my issues. If not, don't. That sounds daunting, right? Not so as you may think. It is important that warning comes earlier so you don't regret what ordinarily you should have avoided.

Let's us put that behind us, but at the back of our mind now that we are about to explain the dynamics of scaling Scrum.

The essence of scaling is to resolve issues associated with agility of the Agile processes and Scrum frameworks. As it stands, there are a number of scaling agile frameworks and they are designed primarily to address these problems. Some of these scaling frameworks include Nexus, Spotify model, Scrum at Scale, Large-Scale Scrum (LeSS), Scaled Agile Framework (SAFe), and Disciplined Agile, among others. Depending on your choice, each of these frameworks works *differently and with a solution in focus.*

Scaling Scrum Frameworks

1. Large-Scale Scrum

Large-Scale Scrum is a framework used for multiple teams for scaling agile development and project. Its working is built on a number of principles that provide simple structures, rules and guidelines. The rules focus on adopting Scrum in large product development

The framework works better and is a perfect starting choice for a team that already has a Scrum in place. It is a great scaling framework for small and medium team. So, if you want to scale up with more teams, then choose LeSS. But the scaling has to be one team at a time.

Features of LeSS

- Works on the mechanism of empiricism, self-managing and self-motivating teams, organizational designs, theory of constraints, systems thinking, lean waste, and queuing theory, etc.
- Provides structures and guidelines for adopting Scrum in big product development.
- Scales up with minimal additional process and not single-team Scrum.
- Perfect for small and medium scrum scaling solutions
- Practices setting agenda to achieve Scrum's purpose

2. Scaled Agile Framework (SAFe)

SAFe is an advanced way to scale Scrum. As an interactive knowledge-based framework, SAFe is designed for large organizations. It is meant to execute agile practices at enterprise or large-scale level. The framework is built with a lot of guidance, covering a wide range of areas including financing and enterprise architecture.

SAFe template seeks to solve organization's issues at four (technically three) levels, namely:

- Team level

- Program level

- Large solution level (optional)
- Portfolio level

At the Team scale, SAFe works like Scrum in conjunction with some Extreme Programming practices. At the level of the Programme, SAFe aligns the team around some additional common events to create an Agile Release Train (ART); the Large solution level is the level where SAFe adds some value stream layer. It is introduced to manage large solutions that can't be handled by a single program. Typically, it is hidden to newbies at SAFe. Meanwhile, SAFe links the goal of the organization to ART at the Portfolio level.

Features of SAFe

- Designed for large sized organizations
- Links strategic enterprise goals to Agile Release Train
- Operates at four levels: three essential, one optional
- Comes with a lot of guidance

3. Nexus

Nexus is a scaling framework intended to integrate teams. It is intended for 3 to 9 Scrum teams. In this sense, a typical Scrum Development Team can make up of 3 to 9 members of team. However, there will be more coordination for bigger teams.

As against other scaling frameworks, Nexus aims to resolve the alignment and integration issues through its Nexus Integration Team. The team has its events which prepare for the individual original Scrum events per team. While it takes its primary role of aligning and integrating, the Nexus Integration Team is also responsible for overseeing the following duties and teams:

- **Nexus Planning** — responsible for discussing the overall scope and dependencies. The discussions from here often led to actualization of Nexus Sprint Goal which is closely followed by Scrum Team also organizing its own events with their individual Sprint Goals in line with the Nexus Sprint Goal.
- **Nexus Daily Scrum** — a meeting preceding Scrum Team Daily Scrums where the team addresses issues bothering on alignment and integration
- **A Nexus Review** — replaces the standard Sprint Review(s).
- A Nexus Retrospective— meets prior to the team's Sprint Retrospective to discuss issues relating to Nexus level.

4. **Disciplined Agile (DA)**

Typically, Disciplined Agile is not designed to be a scaling framework. On the contrary, it is a process decision process model intended to offer a comprehensive guide for an Agile transformation. DA makes use of two of Agile processes including Scrum and Kanban and a host of other transformation knowledge in areas such as Governance, Human Resource and Finance, Portfolio Management and Culture.

As a one stop shop, Disciplined Agile is designed in such a way as to achieve specific goals that allow users to consider their options and learn about their choices. Interestingly, DA is especially attractive for its usefulness because it is designed based on real data. It provides users with an insight into what's going on in other organizations.

As a practice-oriented framework, DA promotes organizational awareness that's based on industry successes, pragmatism, and utility and looks to consider what works and doesn't work in other organization.

Features of DA

- Promotes insight into organizational working and non-functional principles.
- It is a practical framework that operates on the principle of what works and what doesn't work.
- Designed to provide guide for an Agile transformation
- Covers wide range of knowledge based areas like HR, Finance, Portfolio Management, etc.
- Not a scaling framework per se, but a process decision model.

5. Spotify Engineering Culture

The Spotify Engineering Culture is a completely different approach. It comes with a bit of oddity as it offers models that the likes of LeSS, Nexus, SAFe and DA do not.

The Spotify Model does not necessarily utilize Scrum. The team within this framework are at liberty to choose which platform they think is suitable for their project. Meaning that they may or may not go for Scrum.

One other feature is that the model focuses on team autonomy rather than alignment. In other words, the model lays emphasis on team's independence and firm coupling of enterprise structure, design and architecture rather than integrating systems. In that way, there is a severe limit to alignment.

Features

- Offers freedom of framework choice for team
- Focuses on organization system and design
- Emphasizes team autonomy rather than alignment
- It is not necessarily a Scrum model

Chapter 14

CONCLUSION

At the introduction, we give you a story of Jakes and how he was able to overcome the initial difficulty he faced while trying to improve the productivity of his team. Little did Jakes and the Board knew that team, quality delivery and assurance, collaboration and built-in quality are the essentials that the company needed to be effective an efficient. The moment he realized the power of Scrum, then he knew his company was in for a real turnaround. Scrum is indeed the way to go.

www.ingramcontent.com/pod-product-compliance
Lightning Source LLC
Chambersburg PA
CBHW070327220526
45467CB00001B/60